ENDOCRINOLOGY
HORMONES AND HUMAN HEALTH

ENDOCRINOLOGY

HORMONES AND HUMAN HEALTH

PRAKASH S LOHAR

Reader, Department of Zoology
M G S M's Arts Science & Commerce College
Chopda, Jalgaon

MJP PUBLISHERS

Chennai 600 005

Cataloguing-in-Publication Data

Lohar, Prakash S. (1965–   ).
        Endocrinology / by Prakash S. Lohar. –
Chennai : MJP Publishers, 2005.
        xx, 306p. ; 21 cm.
        Includes glossary and index.
        ISBN 81-8094-011-X (pbk.)
        1. Endocrinology I. Title
        573.4   LOH MJP 010

ISBN 81-8094-011-X

MJP PUBLISHERS
A unit of Tamilnadu Book House
47, Nallathambi Street
Triplicane, Chennai 600 005

Publisher : J.C. Pillai
Managing Editor : C. Sajeesh Kumar
Project Coordinator : P. Parvath Radha

Edited and Typeset at ▱▱ Editorial Services, Chennai-5
Cover : R. Shankari   CIP : Prof K. Hariharan

To

Late Dr. S. P. Hardia

FOREWORD

Endocrinology is one of the most fascinating fields of physiology that is playing an important role in human health and welfare. The book will stimulate the interest of young students and kindle in them a desire to read more intensively since it contains every aspect of hormonal coordination in the human body. The integration of cell signalling pathways, involvement of receptors in signal transduction, every aspect of endocrine glands with their dysfunctions, feedback loops and role of hormones in pharmaceutical industry—all these facts of endocrinology are incorporated in a comprehensive form.

The book has been presented in a simple form with appropriate figures that would appeal to the undergraduate and postgraduate students of life sciences including Zoology, Biochemistry, Pharmacy, Health Science, Nursing and Medical Science. It is hoped that the book will meet the need of students who desire to learn clear, up-to-date, and complete instances of endocrine glands that are elaborated in ten chapters. The presentation of every hormonal disorder with its laboratory evaluation, related symptoms, and illustrations based on a combination of quality and adequacy are significant markers of this book.

I feel proud and happy at the outcome of the member of our association (IAAB) in putting together a coherent picture of endocrinology. The efforts made by Dr. Lohar to organize the text matter systematically are highly appreciable. The author with his long teaching and research experience has done justice to the subject. I am sure the book will generate enough interest in the minds of brilliant students who may pursue the subject in his or her research career. The book will be very much useful for students as well as teachers pursuing a career in teaching, research or extension fields related with medicine, veterinary science, fishery science and other biological sciences.

Mohan S. Kodarkar
Secretary
Indian Association of Aquatic Biology
P.G. Department of Zoology
Vivek Vardhini College
Hyderabad (A.P.)

PREFACE

Endocrinology: Hormones and Human Health is the book for the graduate and postgraduate students of Zoology, Biochemistry, Pharmacy, Health Science, and Medical Science. The purpose of writing this book is to elaborate fundamental concepts as well as current thrusts in the field of regulatory mechanisms in the human body, which are principally based on hormonal coordination and applicable to all mammals. I have tried to make this text interesting and readable for the relevant students and teachers. The book contains many more facts and details presented in a simplified language that a student either needs or should attempt to learn.

In all, the text has been suitably divided into ten chapters. Brief descriptions of various aspects of hormones, cell signalling, signal transduction, role of cAMP, cGMP, and other second messengers that are derived form phosphatidylinisitol, the involvement of receptor tyrosine kinase, protein kinase A and trimeric G protein have been explained in the first chapter. The detailed discussion of all endocrine glands, their relationship with hypothalamus and pituitary gland and the physiological role of various hormones are included in chapters 2 to 8. The hormonal deficiency or its excessive secretion leading to the relative disorders in human beings is the highlight of this publication. The thyroid function test, glucose tolerance test, and the significance of insulin-like growth factors are also elaborated in relevant chapters.

The feedback mechanisms with interesting examples are illustrated in chapter 9. Chapter 10 is purposefully included in this book to enlighten the students about the role of hormones as pharmaceuticals. It also deals with the biotechnological approach for the synthesis of hormones on a commercial basis from the genetically modified organisms. At the end of each chapter there is a synopsis and review questions that are designed to identify key concepts and review the material to test their comprehension respectively. Every attempt has been made to update all chapters. Most of the illustrations of endocrine glands given in the text are diagrammatic, and are not intended to be anatomically accurate. I accept the responsibility for errors of facts and judgement.

Prakash S Lohar

ACKNOWLEDGEMENTS

I offer my grateful thanks to Dr. M. S. Kodarkar, Secretary, Indian Association of Aquatic Biologists (IAAB), Hyderabad for writing the foreword for this book.

I am deeply indebted to honorable Dr. S. G. Patil (Gynaecologist and Chairman, MGTSM's A. S. C. College, Chopda), Dr. P. M. Mahajan (Pathologist), Dr. M. T. Jaiswal (Gynaecologist), Dr. Nachiket Potdar and Dr. Swati Chavan, who have continuously encouraged me to write this book and have provided valuable facts and opinions about the hormonal disorders and drug therapy.

My special gratitude is due to Dr. P. S. Hardia (An Eye Surgeon from Indore, who paved his way into the Guinness Book of World Records) for his cooperation with my family in an extraordinary way and for providing access to internet and superb medical library whenever I stayed in his research institute. I wish to express my sincere thanks to distinguished authors of related books and articles on web pages, which is an essential prerequisite to writing.

I would like to express my thanks to my mother, wife, and my kids as well as to Shri. P. N. Lohar, Nitin Lohar, Sharad Lohar and Dr. B. B. Waykar (Rashtriya College, Chalisgaon) for their moral support throughout the preparation of this manuscript. Thanks are also due to the Management, the Principal, and Vice principal of my college, and to my friends and colleagues for their constant encouragement.

Finally, I sincerely acknowledge the cooperation and determination of Mr. Pillai, Mr. C. Sajeesh Kumar and other staff of MJP publishers, Chennai for producing this project on time.

I cordially invite all the readers to send their useful comments and criticism.

Prakash S Lohar

CONTENTS

ONE

INTRODUCTION

1.1 CONCEPT OF SECRETION

Endocrinology is the science that deals with transfer of information by chemical integration within an organism. Communication within cells, tissues, organs and various systems is essential for coordinating the physiological processes and the happenings in the external environment. The nervous system is an exceedingly complicated but highly efficient method of coordinating the activities of the body by 'messages' received from and sent to specific areas of the body with speed and precision. There is a second integrative system in the body, in which chemical 'messengers' called *hormones* produced by *endocrine glands* carry the information and directives through the blood (figure 1.1).

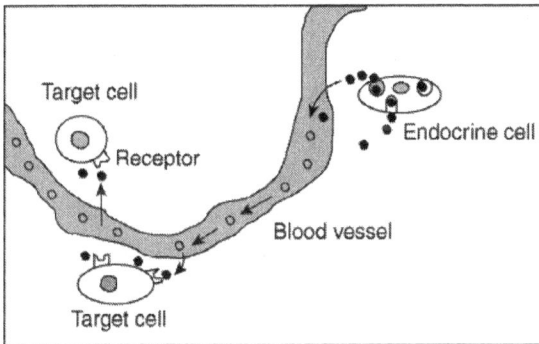

Figure 1.1 Hormones circulated by blood, bind to receptors on distally placed target cells.

Secretory activity—the production and discharge of fluid material—is the characteristic of cells. The gland cells are the epithelial cells that do this job predominantly; in addition nerve tissue may also secrete substances called *neurohormones*. Commonly a secretory activity may be carried on indefinitely without apparent harm to the cell; this is the condition in *merocrine* type of gland. But if the secretion is thick or viscous, the cell usually is damaged while discharging it. In *apocrine* type, the superficial part of the cell breaks down as the secretion is released, but no permanent harm is done; this part of the cell is reconstituted and secretion recurs. In *holocrine* type of secretion the entire cell is destroyed with the discharge of its content as in sebaceous glands in mammalian skin.

Unlike the endocrine cells, there are cells which are not assembled into glands but secrete certain regulatory proteins called growth factors that attach to the external cellular receptors and stimulate the growth and differentiation of cells. Growth factors were first discovered when scientists attempted to culture cells outside the body. Even when given all sort of nutrients and optimal conditions, the cells did not grow well unless blood plasma or tissue extract was added to the medium. The components necessary for growth were found to be specific molecules that were present in very small quantities. At present, about 50 specific growth factors are known. Some of them are PDGF (platelet-derived growth factor), EGF (epidermal growth factor), CSFs (colony stimulating factors) and interlukins. Most of the growth factors are *paracrine* in their action—they diffuse only short distances and act only on local cells, while some are *autocrine* (self-stimulatory) in action that is, they act on the cells secreting them (figure 1.2). When the primary effects of a hormone are on cells near its site of release, the hormone is said to have paracrine function. Paracrine hormones are released in such tiny quantities, or are so rapidly inactivated by degradative enzymes, or are taken up so efficiently by local cells that they never diffuse into the blood in sufficient amounts to act on distant target cells. An example of a paracrine hormone is histamine that mediates the process of inflammation. Nerve cells, or neurons, can also be considered as paracrine cells. A neuron communicates with another cell by means of a chemical messenger called a neurotransmitter, which travels over a small distance to the target cell.

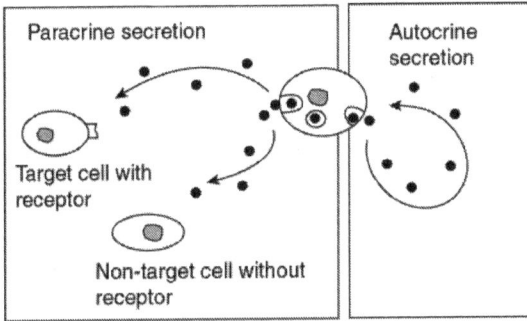

Figure 1.2 An autocrine hormone is self-stimulatory; a paracrine hormone influences nearby cells.

When a hormone binds to receptors on the same cell that is releasing it, the hormone acts as an autocrine message. Autocrine pattern of secretion is executed by interlukins 2 (growth factor secreted by T cells) that stimulates proliferation of T lymphocytes and also by many human lung cancer cell lines that secrete proteins belonging to the family of bombesin-like peptides (BLPs), and these cells also have receptors for BLPs. Bombesin was originally isolated from the skin of the frog *Bombina bombina*. It is a peptide that can stimulate pancreatic enzyme secretion. Experimental evidences proved that monoclonal antibodies that bind to bombesin and block its binding to cellular receptors can inhibit the growth of human lung cancer cells growing *in vitro* or *in vivo* following their subcutaneous injection into mice. These results support the autocrine hypothesis of tumour formation.

1.2 HORMONES AS MESSENGERS

The glands present in the body may be classified into *exocrine* glands that pass their secretion along the ducts to the external surface of the body like sweat, sebaceous, salivary, mammary glands, etc. and *endocrine* or ductless glands that pour their secretions called hormones, directly into the bloodstream or lymphatics. Hormones are the special *chemical messengers* that are carried to their respective target organ(s) where they initiate changes or activities in them that may last for a few seconds, few days, months or even years, which ultimately help the organism to adjust to the new situation. Thus hormones are the products of ductless glands that have specific physiological functions, usually elsewhere in the body.

In 1909 Bayliss and Starling introduced the term hormone for the chemical messenger that is released from one group of cells and travel via blood stream to affect one or more different group of cells. Huxley in 1935 defined hormones as "information transferring molecules, the essential function of which is to transfer information from one set of cells to another, for the good of the cell population as a whole". According to Selye (1948) hormones are "physiological and organic compounds produced by certain cells for the sole purpose of directing the activities of distant parts of the same organism".

In addition to hormones there are neurotransmitters or neurohormones secreted by modified motor neurons. They act on other neurons, muscles or glands that are in close contact with the nerve endings and produce local and more rapid action. Neurotransmitters are then rapidly destroyed enzymatically near the site of action. For coordination and integration of the various body activities, both neurohormone and hormone are very essential and often one supplements the function of the other.

1.3 CLASSIFICATION OF HORMONES

The classification of hormones is determined by their origin and not by the type of their function. Chemically, three categories of hormones may be recognised:

1.3.1 Steroid Hormones

These hormones are derived from cholesterol. They are lipid-soluble and can easily dissolve in and pass through cell membranes. Therefore, steroid hormones are not packaged in vesicles; instead, they simply diffuse out of the cells that make them as they are synthesized. The different steroid hormones are:

i.   Adrenocortical steroids—mineralocorticoids, aldosterone, deoxycorticosterone, glucocorticoids, cortisone, cortisol and corticosterone.

ii. Ovarian hormones—β-estradiol, estriol, estrone and progesterone (also secreted by placenta, adrenal cortex and testes).

iii. Testicular hormones—testosterone, androsterone and dehydroepiandrosterone.

1.3.2 Protein or Peptide Hormones

There are number of endocrine organs which secrete hormones that are chemically made of proteins or large polypeptides. They are water-soluble and therefore easily transported in the blood, but they cannot easily pass through lipid-rich cell membranes. Therefore, peptides and protein hormones are packaged in vesicles in the cells that make them, and released by exocytosis. A few examples are given below.

i.   Hormones secreted by hypothalamic nuclei—oxytocin (pitocin), vasopressin (antidiuretic hormone)

ii. Hormones of adenohypophysis—somatotropin (STH), thyrotropin (TSH), corticotropin (ACTH), and gonadotropin (GTH—FSH, LH, LTH).

iii. Melanocyte-stimulating hormones (α and β) secreted by pars intermedia.

iv. Insulin and glucagon secreted by islets of Langerhans in pancreas.

v.   Calcitonin secreted by thyroid and parathormone released from parathyroid gland.

vi. Hormones of gastrointestinal tract—parotin, gastrin, secretin, cholecystokinin, pancreozymin, hepatocrinin, enterogastrone, enterocrinin and villikinin.

vii. Relaxin secreted by ovaries, placenta and uterus.

1.3.3 Amino Acid Derivatives

Two important groups of hormones are the derivatives of the amino acids, tyrosine and phenylalanine. Some amine hormones are water-soluble and others are lipid-soluble; thus, their mode of release differs accordingly.

i.   Thyroxine and triiodothyronin are iodinated derivatives of tyrosine secreted by thyroid gland.

ii. Catecholamines called adrenaline and noradrenaline produced by adrenal medulla, both derived from phenylalanine.

1.4 CHARACTERISTIC FEATURES OF HORMONES

1. Hormones are secreted by endocrine glands.
2. They act on target organs that may be distally located.
3. Hormones show high degree of specificity.
4. They are secreted in response to specific stimuli.
5. Hormones are soluble in water, hence easily transported via blood.
6. They are non-antigenic.
7. Most of the hormones have low molecular weight and therefore they can easily pass out through blood capillaries.
8. They act in very low concentration but their effect is long-lasting.
9. Once their function is over, hormones are readily destroyed by enzymatic action or are inactivated or excreted.
10. Hormones are not species-specific because hormones extracted from animals are found to be effective in man.

1.5 DISCOVERY OF HORMONES

Some of the important events in relation to the science of endocrinology are given below.

Aristotle (382–384 BC)	described the effect of castration in men and birds.
Ruysch (1690)	described the production of an important substance by thyroid that entered the blood stream.
Berthold (1849)	performed the testicular grafting to induce comb growth in a capon.

Claude Bernard (1855)	coined the word 'internal secretion' and demonstrated that the liver releases sugar directly into blood.
Thomas Addison (1855)	demonstrated the existence of human syndrome (now called Addison's syndrome) associated with deterioration of adrenal cortex.
Baumann (1895)	observed high concentration of iodine in thyroid gland.
Brown Sequard (1899)	noted the improvement in his sexual activity after taking cutaneous injection of aqueous extract of dog testis at the age of 72.
Mering and Minkowski (1899)	produced diabetes mellitus in dog after surgical removal of the pancreas.
Bayliss and Starling (1909)	introduced the term 'hormone' and 'secretin' was the first one to be recognised by a name.
Earl Sutherland (mid-1950's)	discovered the second messenger in the form of cyclic AMP that initiates the binding of hormone to plasma membrane and changes the activities of enzymes.
F. Sanger (1958)	discovered the primary structure of the hormone insulin.
C. B. Huggins (1966)	elucidated hormonal treatment for prostatic cancer.
E. W. Jr. Sutherland (1971)	discovered the mechanism of hormonal action.
R. Guillemin and A. V. Schally (1977)	noted the production of peptide hormone by the brain.
A. G. Gilman and M. Rodbell (1994)	discovered the role of G-proteins in signal transduction in cells.
G. Blobel (1999)	discovered that proteins (and peptide hormones) have intrinsic signals that govern their transport and localization in the cells.

1.6 MECHANISM OF HORMONE ACTION

Hormones act on every major group of tissues in the body. The way in which the different hormones control activity levels of target tissues differs from one hormone to another. In most of the cases, they act by regulating preexisting patterns of cellular reactions and not by themselves acting as enzyme or coenzyme. It is generally considered that a hormone, in order to act, must first bind to specific receptors on their effector cells, the union of which then triggers a chain of events that dramatically change the preexisting patterns for synthesis of enzyme or function. There are two important general mechanisms by which hormones act on the cells.

1. Activation of the cyclic AMP system of the cells, which in turn elicits specific cellular function, or

2. regulation of the genes in the nucleus of the cell which then synthesize protein or enzyme that in turn initiates the cellular response.

1.7 CELL SIGNALLING

Communication between cells depends on the release of signalling molecules from the cells that migrate to other cells (target cell) and deliver stimuli to those equipped to receive the signals. Regardless of the nature of the signalling, the molecule that binds to the receptor at the outer cell surface is called *ligand* or *agonist;* it has a precise relationship to its receptor as a substrate has to its enzyme.

In order to survive, it is essential for the cells to communicate with their neighbours, to monitor the conditions in their environment and to respond according to the type of the stimuli. These interactions are effected by a phenomenon called *cell signalling*, by which information is relayed across the plasma membrane to the cytoplasm and even to the nucleus of the cell (figure 1.3). The steps involved in cell signalling are as follows:

1. The stimulus is detected at the outer surface of the plasma membrane by a specific receptor present within the cell membrane.

2. Signal is transferred across the cell membrane to its cytoplasmic surface.

3. Receptors present on the inner surface of the membrane or within the cytoplasm help in transmission of the signals to trigger the cellular response.

4. Steroid hormones and thyroxine, being lipid-soluble, are transferred through the lipid bilayer of the cell without difficulty and their receptors are located inside the cell rather than on the membrane.

5. Water-soluble hormones cannot pass through the lipid bilayer and hence they bind to membrane receptors.

6. The different cellular responses arising due to binding of signalling molecules (whether they are water-soluble or lipid-soluble hormones) with the receptors may include activation of the enzyme, a change in ion permeability, a reorganization of cytoskeleton, or even the activation of gene for the synthesis of DNA or RNA.

7. Finally, the cell ceases to show the response due to inactivation or destruction of the signalling molecule.

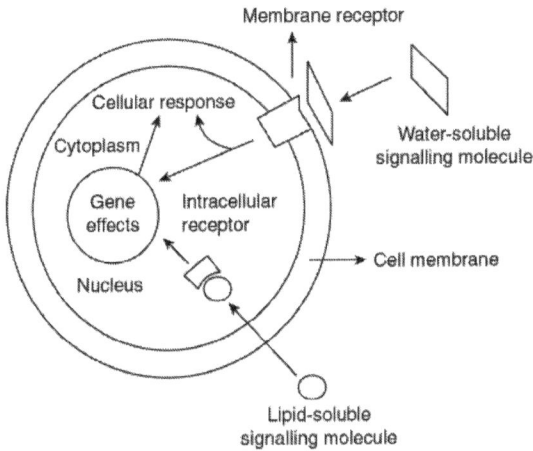

Figure 1.3 General view of receptor-mediated signalling of the cell.

1.8 SIGNAL TRANSDUCTION

It is another term commonly used in connection with cell signalling that indicates the nature of the stimulus received by the cell-surface receptor and is entirely different from the signal that is released to the internal compartment of the cell. The plasma membrane plays a crucial role in showing the response of the cell to external stimuli, the process called as signal transduction. It is a complex process that involves transfer of information along the signal transduction pathways having the machinery for physically transporting substances from one side of the membrane to another with the help of a chain of distinct proteins present in the plasma membrane. The passage of information through the cell is similar to the passage of electrons along the electron transport chain. In both cases, each component of the pathway is altered by the element lying 'upstream' in the pathway and, in turn, alters the element lying 'downstream' in the pathway. Signal transduction pathways consist primarily of protein kinases and protein phosphatases— the enzymes whose catalytic actions change the confirmations of the next protein in the series, further activating or inhibiting the protein to show specific response. Some of the kinases and phosphatases have numerous proteins as their substrates, whereas others phosphorylate or dephosphorylate only a single protein substrate. Many of the substrates of these enzymes are other enzymes, but when kinases and phosphatases act on protein substrates that may include ion channels, transcription factors, and various types of regulatory subunits, the addition or removal of the phosphate groups from the substrate triggers a conformational change in the protein that activates or suppresses its activity, thereby showing specific response (figure 1.4).

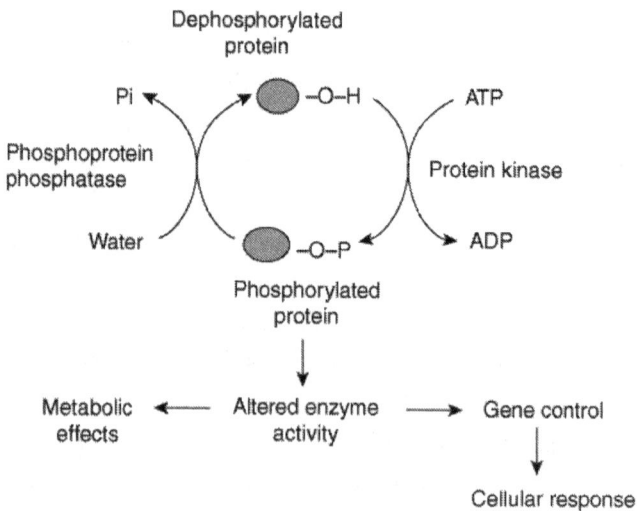

Figure 1.4 Signal transduction through protein kinase and phosphoprotein phosphatase.

1.9 CYCLIC AMP—A SECOND MESSENGER

Earl Sutherland and his colleagues at Case Western Reverse University, and Edwin Krebs and Edmond Fischer at the University of Washington, in the mid-1950s, elucidated the connection between the binding of a hormone to the plasma membrane and the change in the activities of phosphorylase and glycogen synthase. A wide variety of hormones (acting as first messengers) use cAMP as *second messenger*, a substance that is released into the interior of the cell as a result of the binding of a first messenger—a hormone or other ligand—to a receptor at the outer surface of the cell. Cyclic adenosine monophosphate (cyclic AMP, or simply cAMP) is produced from ATP by an enzyme, adenylate cyclase that is attached to the inside of the cell membrane (figure 1.5).

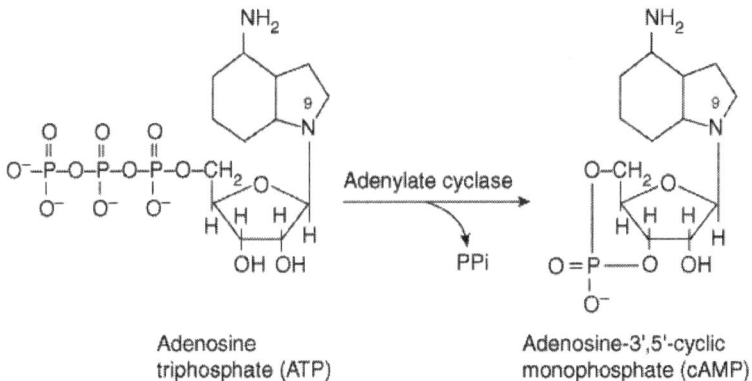

Figure 1.5 The formation of cyclic AMP from ATP catalysed by adenylate cyclase

The cascade of reactions involved in the production of glucose by liver cells under the influence of glucagon or epinephrine mediated by cAMP is illustrated in figure 1.6.

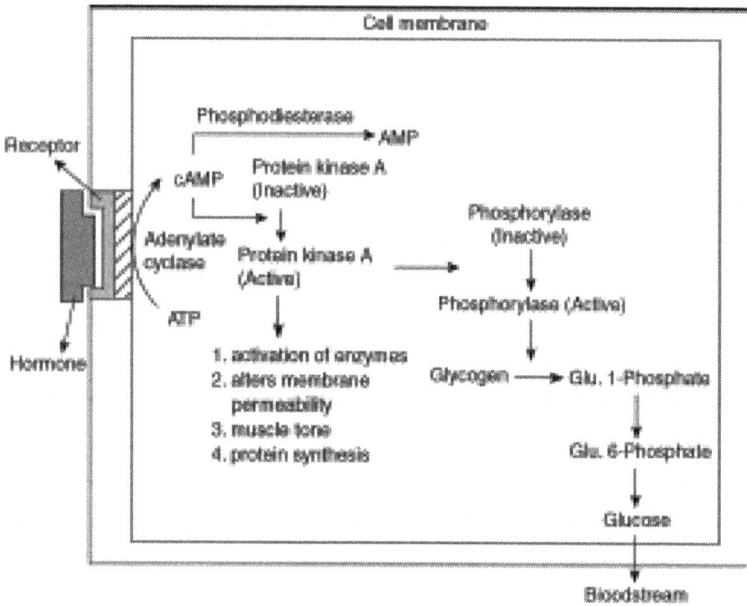

Figure 1.6   Diagrammatic representation of the role of cyclic AMP as a 'second messenger' in the mechanism of hormone action.

The hormone glucagon or epinephrine (first messenger) binds to the outer surface of a liver cell that causes a change in the conformation of the receptor, and is transmitted across the plasma membrane and activates the adenylate cyclase at the membrane's inner surface. On activation, adenylate cyclase catalyses the conversion of ATP into cyclic AMP (second messenger) that can rapidly diffuse into the cytoplasm where it evokes a response by triggering a chain of reactions. Each cAMP then activates the enzyme protein kinase A (PKA) which is able to phosphorylate another enzyme phosphorylase kinase, that in turn catalyses the breakdown of glycogen to glucose-1-phosphate and finally to glucose, which diffuses into blood and reaches the tissues of the body.

The cAMP mechanism is executed in different cells in response to a wide variety of hormones to act as an intracellular hormonal mediator system. Some of the responses mediated by cAMP in mammalian cells are as follows:

- Hydrolysis of glycogen, glucose synthesis in liver cells due to epinephrine and glucagon
- Increased contractility of skeletal muscles by epinephrine
- Secretion of thyroid hormones under the impact of TSH
- Increased permeability of nephrons in kidney due to ADH
- Control over extracellular calcium by parathyroid hormone
- Secretion of steroid hormones from ovary due to LH
- Increased secretion of glucocorticoids from adrenal cortex under the influence of ACTH
- Control over the contraction of many smooth muscles by epinephrine
- Catabolism of triacylglycerol in adipose tissue by epinephrine, glucagon and ACTH
- Secretion of hypothalamic releasing factors that in turn control secretion of most of the hormones from the anterior pituitary

1.10 CYCLIC AMP-MEDIATED SIGNAL TRANSDUCTION FOR ACTIVATION OF GENES

It has been described how cAMP is produced in cytoplasm and in turn activates the PKA (protein kinase A). While most of the activated PKA molecules remain in the cytoplasm, some of them play a significant role in activation of genes. PKA is a part of an important pathway of gene control. A fraction of cytoplasmic PKA translocates into nucleus where it phosphorylates nuclear proteins called CREB (cAMP-responsive element binding protein). The phosphorylated CREB becomes an active transcriptional factor that binds to the DNA having a particular sequence of nucleotides that are known as the cAMP-regulated enhancer (CRE). The binding of phosphorylated CREB and CRE on DNA initiates the transcription and synthesis of mRNA that in turn is translated to form anabolic enzymes (figure 1.7). Thus, when the liver cell shows response to glucagon or epinephrine, it activates not only the catabolic enzymes that are required for breakdown of glycogen but also the genes containing CREs for the synthesis of anabolic enzymes required to form glucose from non-carbohydrate precursors (the process is called gluconeogenesis).

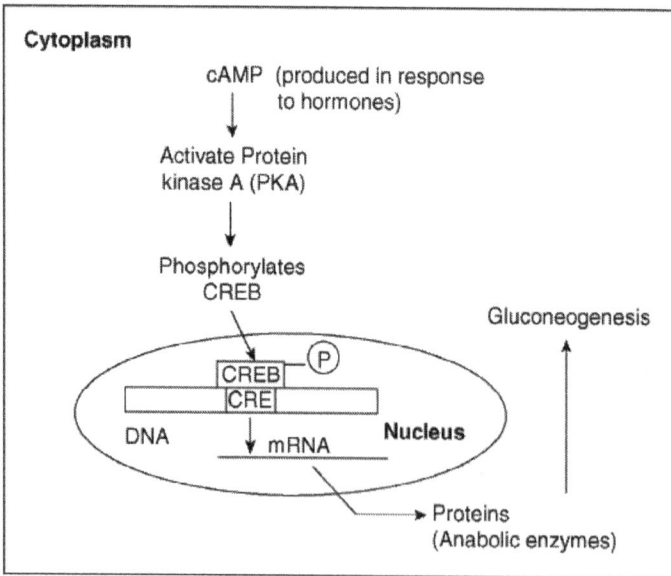

Figure 1.7 Cyclic AMP-mediated activation of gene by phosphorylated CREB.

1.11 ROLE OF G-PROTEIN IN SIGNAL TRANSDUCTION

The effect of hormone receptor complexes on adenylate cyclase is indirect. It is mediated by a third membrane protein called *GTP-binding protein*, or *G protein*. This protein binds guanine nucleotide, GDP, on a cytoplasmic face until it collides with a specific hormone-receptor complex. Then it is activated, exchanges its GDP for GTP, and as a result binds to and alters the activity of adenylate cyclase. The reaction terminates when GTPase hydrolyses the GTP back to GDP. Over 100 different G-protein-coupled receptors have been identified that respond to a wide variety of stimuli. All these receptors act through a similar mechanism.

A regulatory G protein is associated with the cytoplasmic face of the receptor protein. It has three subunits, α, β, and γ. Since these three polypeptides are different, it is referred to as a *heterotrimeric G protein* or simply, *trimeric G protein*. Each G protein can exist in two states: an active state with bound GTP or an inactive state with bound GDP. The α subunit of the G protein has a site on it that can bind to GDP or GTP. In the absence of the hormone, the site on the α subunit is occupied by GDP. When a molecule of the hormone binds to its specific receptor, it causes a change in the conformation of the receptor that increases its affinity for G protein. As a result, the hormone-bound receptor binds to the G protein, causing it to release its bound GDP and bind a GTP that switches the G protein into active state. An exchange of guanine nucleotides alters the conformation of the G_α subunit, causing it to dissociate from other two subunits that remain together as a $G_{\beta\gamma}$ complex. The dissociated G_α subunit forms the complex with GTP and further activates an adenylate cyclase to produce cAMP. The complex of the α

subunit of the G protein and GTP acts as an enzyme, GTPase, that hydrolyses GTP to GDP and Pi, which shuts off the subunit's ability to activate additional adenylate cyclase. The G_α – GDP then reassembles with the $G_{\beta\gamma}$ subunits to reform the original trimeric complex and thus the system returns to its resting state (figure 1.8).

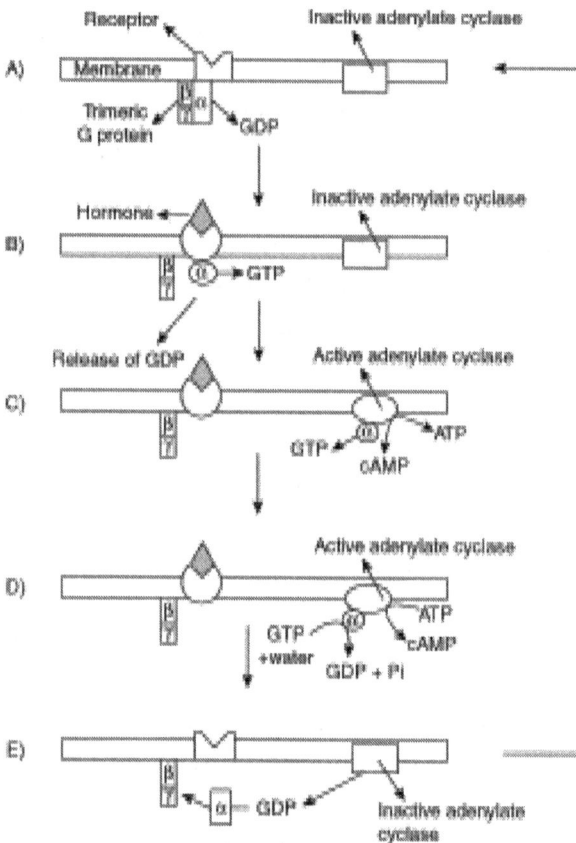

Figure 1.8 Mechanism of receptor-mediated activation or inactivation of adenylate cyclase via heterotrimeric G protein.

The GTP hydrolysis that inactivates adenylate cyclase can be exemplified in the infection of the bacterium *Vibrio cholerae* to cause cholera. The cholera toxin inactivates the GTPase activity of G_α subunit in the cells of the intestinal epithelium. As a result, adenylate cyclase remains in an active mode causing prolonged cAMP production resulting in massive Na^+ loss, accompanied by large volumes of intestinal fluid to be secreted into intestinal lumen. The loss of water and electrolyte from the body (diarrhoea) may often lead to death. In addition, there are a number of inherited disorders that have been traced to defects in cell-surface receptors or their defective G protein. Congenital nephrogenic diabetes insipidus (CNDI) is one of such inherited diseases in which infants suffer from serious dehydration as a result of the inability to produce concentrated urine. If not diagnosed promptly, chronic dehydration can produce mental retardation, inadequate growth, and even death. The disorder is due to the presence of mutated G protein in the cells of kidney to make them unresponsive to the hormone vasopressin (antidiuretic hormone).

1.12 CYCLIC GMP AS SECOND MESSENGER IN SIGNAL TRANSDUCTION

There is a diverse group of hormones that use the cyclic nucleotides, cAMP and cGMP as second messenger. Cyclic GMP is analogous to cAMP, which can be formed by the action of guanylate cyclase enzyme from GTP (figure 1.9).

Cyclic GMP mediates the cell response by activating specific protein kinases. The effects produced by cGMP are more specialized than those of cAMP. Relaxation of smooth muscles in blood vessels of heart, effects on nerve cells and

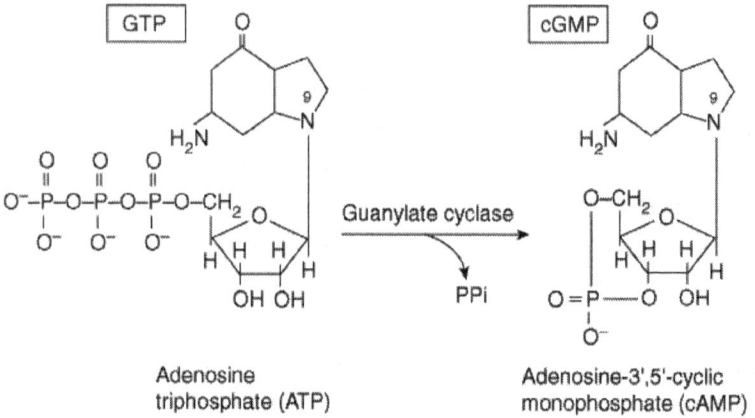

Figure 1.9   Formation of cyclic GMP by guanylate cyclase.

vision are mediated by cGMP. On the basis of experimental evidences it was discovered that the enzyme guanylate cyclase is activated by a substance, NO (nitric oxide) that acts as intracellular messenger stimulating the cell's phagocytic activity. There are a number of other physiological activities in which NO is turning up as a messenger. Production of nitric oxide by the endothelial cells that are present in the inner surface of the blood vessels in response to the neurotransmitter acetylcholine to mediate the nervous control. The formation of acetylcholine–receptor complex in the plasma membrane of endothelial cells activates the enzyme nitric oxide synthase that generates NO from arginine guanidino substrate. NO easily diffuses from the cells producing it and enters the adjacent smooth muscle cells where it stimulates the enzyme guanylate cyclase to produce cGMP which in turn triggers the response leading to muscle relaxation and dilation of blood vessel (figure 1.10).

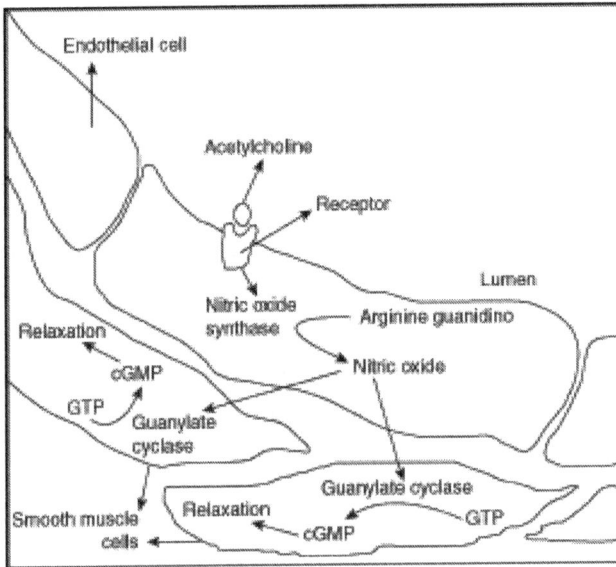

Figure 1.10 Involvement of NO and cGMP in signal transduction for vasodilatation.

1.13 ROLE OF CALCIUM–CALMODULIN SYSTEM IN CELL SIGNALLING

There are various physiological activities like cellular movement, endocytosis, metabolism, secretion, fertilization, synaptic transmission, and cell division where the calcium ions play an important role. With the help of computerized imaging technique employing fluorescent calcium-binding dyes to observe the concentration of free calcium ions in the cytoplasm, it has been revealed that there is dramatic change in the concentration of free cystolic calcium. It also shows that the level of free calcium ions fluctuates as per the type of the cell and the type of stimulus. Although the calcium ion is very different in structure from cAMP or cGMP and it is not a substance that is synthesized by enzyme action, it does share an important property with other cytoplasmic messengers. Calcium does not act on different targets in free ionic state, but instead its role as a second messenger is associated with a widely distributed protein known as calmodulin, a small polypeptide discovered in the late 1960s in the tissues of brain. Each molecule of calmodulin has four sites having high affinity for binding Ca^{2+} ions. The binding of one or more calcium ions causes a conformational change in calmodulin. On binding with Ca^{2+} ions, the change in the conformation of calmodulin alters the activity of its associated enzyme. The Ca^{2+}–calmodulin complex can initiate variety of cellular responses as per the type of cell by activating protein kinase, a cyclic nucleotide phosphodiesterase or calcium-transport system of the cell membrane. Figure 1.11 shows the mechanism of smooth muscle contraction in which the calcium–calmodulin complex

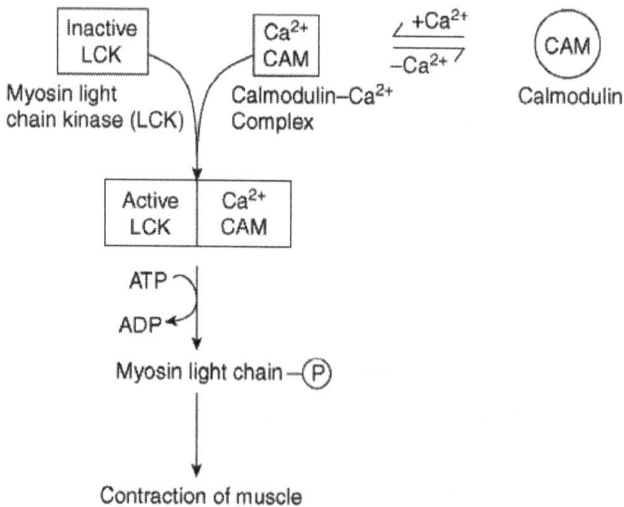

Figure 1.11   Mechanism of smooth muscle contraction involving Ca^{2+}–calmodulin system.

plays a key role in the activation of myosin kinase. The phosphorylation of myosin light chain triggers the contraction of smooth muscle.

1.14 SIGNAL TRANSDUCTION USING RECEPTOR TYROSINE KINASES (RTKS)

It is evident that a wide variety of hormones trigger a wide variety of cellular responses mediated by different messengers. One of such hormones is insulin that is secreted by the β cells of the islets of Langerhans in pancreas. It serves as a hypoglycemic factor by removing glucose from blood and polymerising it into glycogen by acting on the liver and muscles. In addition to its profound effect on carbohydrate metabolism, insulin also promotes lipid and protein synthesis, as well as the growth and proliferation of cells. Research has been made in this direction to find out the exact mechanism of signalling pathway for insulin and in what respect it differs from the other pathways used by hormones like glucagon and epinephrine. The liver and muscle cells have the ability to respond to insulin due to the presence of insulin receptor in their plasma membrane. The insulin receptor is not just a protein that binds with a ligand but also an enzyme—a protein tyrosine kinase that phosphorylates the tyrosine residue present in the other proteins (figure 1.12).

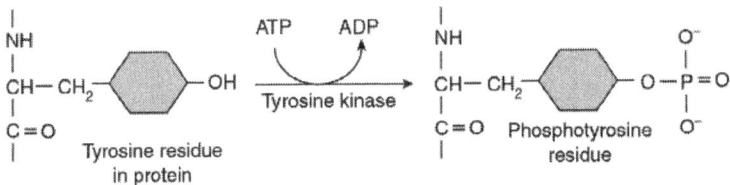

Figure 1.12 Phosphorylation of tyrosine residue by tyrosine kinase.

Since insulin receptor has enzyme activity, it is known as receptor tyrosine kinase (RTK). Extensive research in this field has lead to the identification of over 50 different RTKs and their number is ever-increasing. The insulin receptor has two α and two β polypeptides in its structure constituting a tetrameric protein. The α-chains of the receptor having insulin-binding sites are present on the external surface of the plasma membrane while the β-chains extend across the membrane and convey the signal to its inner surface at which tyrosine kinase domain is present. When the bound insulin is absent, the tyrosine kinase function of the receptor is inactive; the binding of insulin to the α-chains of the receptor causes alteration in the conformation of the β-chains of the receptor and subsequently activates the tyrosine kinase. Activation of tyrosine kinase further catalyses the transfer of phosphate groups to tyrosine residues that are located in the cytoplasm domain of the receptor (autophoshorylation) and to a variety of insulin receptor substrates (IRSs). The phoshorylation of the IRS proteins in turn activate the downstream signalling pathways like a) stimulation of the movement of glucose transporters to the plasma membrane to import additional glucose molecules, b) activation of the transcriptional factors for expressing insulin-specific genes, c) promoting DNA synthesis and further cell division, or d) activating genes for protein synthesis (figure 1.13).

The RTKs are present on the surfaces of the specific target cells and have the ability to interact with a wide variety of extracellular agents including insulin and can regulate diverse functions like cell growth and multiplication, the course of cell differentiation, the uptake of foreign particles by the process of phygocytosis, and the survival of the cell.

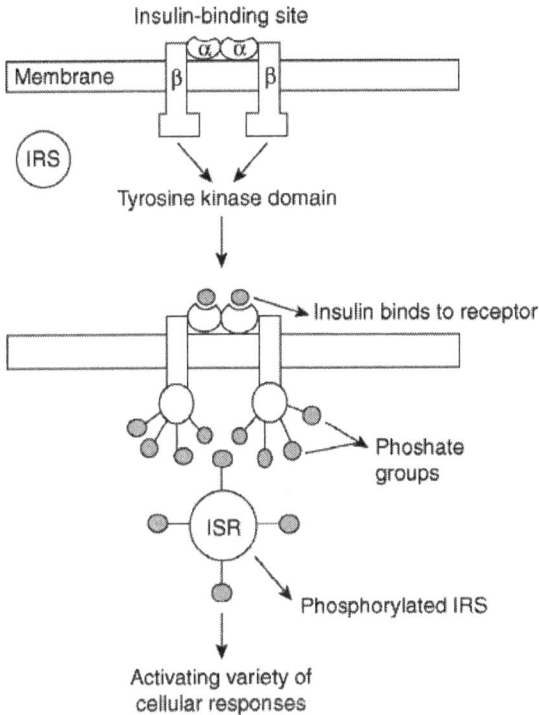

Figure 1.13   Response of tetrameric receptor to insulin, leading to activation of cellular functions via autophosphorylation and phosphorylated IRS.

1.15 CELL SIGNALLING BY PHOSPHATIDYLINISITOL-DERIVED SECOND MESSENGERS

The generation of the second messengers that are derived from phosphatidylinisitol forms one of the signal transduction pathways, which may be activated by G protein-coupled receptors. Phosphatidylinisitol (PI) is present in the plasma membrane as one of structural components of lipid bilayer and acts as the precursor for the other key regulatory molecules that are triggering specific cellular responses. A single phosphate group added to inisitol sugar of PI generates PI 4-phosphate (PIP) that can be phosphorylated further to form PI 4, 5-diphosphate (PIP_2). When the neurotransmitter acetylcholine binds to the surface of a smooth muscle cell lining of a blood vessel to show the response in the form of vasoconstriction or when an antigen binds to a mast cell to trigger the cellular response in the form of secretion of histamine for allergic attack, the bound receptor activates a heterotrimeric G protein, which in turn stimulates an enzyme known as phospolipase C—the effector present in the inner surface of the plasma membrane. This enzyme further catalyses breakdown of PIP_2 into inisitol 1,4,5-triphosphate (IP_3) and diacylglycerol (DAG). Both these molecules act as second messengers in signal transduction pathway.

IP_3 is a small, water-soluble molecule that rapidly diffuses from membrane to cytoplasm where it binds to a specific IP_3 receptor present on the surface of the smooth endoplasmic reticulum (stores Ca^{2+} in most of the cells) and allows Ca^{2+} ions to diffuse into the cytoplasm. Further, its binding to various target molecules initiates specific cellular responses (figure 1.14). In both the above-mentioned examples, the contraction of smooth muscle in blood vessel and secretion of histamine by mast cell, the responses are triggered due to increased levels of calcium.

As stated earlier, DAG is formed by the catalytic action of phospolipase C enzyme. It is a lipid molecule which remains in the cell membrane after its formation where it activates an enzyme protein kinase C. Similar to the enzyme protein kinase A that is activated by cAMP, protein kinase C has a significant role to play in cellular activities.

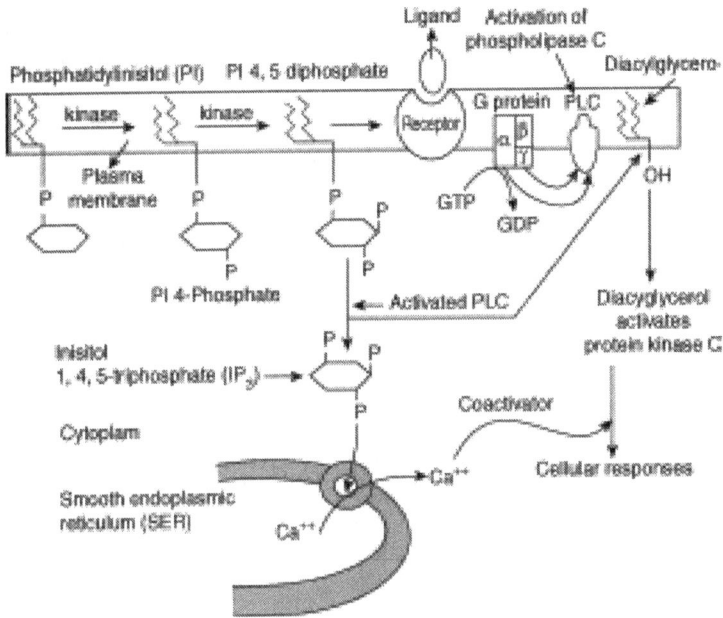

Figure 1.14 Activation of phospolipase C mediated by G protein to produce Phosphatidylinisitol-derived second messengers, IP_3 and DAG.

1.16 STEROID HORMONE RECEPTORS

As mentioned earlier and shown in figure 1.3, steroid hormones, for example glucocorticoids, aldosterone, estrogen, progesterone, etc. are lipid soluble ligands that easily pass through the lipid bilayer of cell membrane and their receptors are present inside the cell rather than on the plasma membrane. The receptor for glucocorticoid is located in the cytoplasm while sex steroid receptors are present in the nucleus. There are evidences regarding the ultimate impact of most of the steroid hormones to increase the growth, development and differentiation of tissues and this is associated with increase in the rate of protein synthesis or the formation of the enzymes for performing specific cellular function. These facts confirm the regulatory role of steroid hormones on the genes through transcriptional factors.

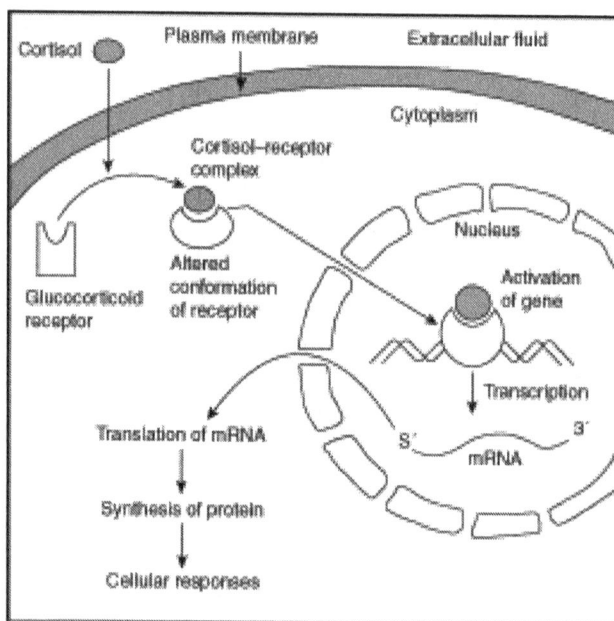

Figure 1.15 Mode of action of steroid hormone mediated by intracellular receptor.

Glucocorticoids (such as cortisol) are steroid hormones, secreted by the middle zone of the adrenal cortex promoting the conversion of amino acids to glucose and its uptake by the brain. These hormones are secreted in large quantities during the periods of stress that may be caused by starvation or severe physical injury. The cells responding to glucocorticoids must have a specific receptor with the ability to bind the hormone. Cortisol enters the cell from the extracellular fluid diffusing through the lipid bilayer and into the cytoplasm, where it binds to a glucocorticoid receptor. The formation of cortisol–receptor complex alters the conformation of the receptor and subsequently translocates into nucleus, where it acts as a transcriptional factor and binds to glucocorticoid-responsive elements or sequences of the DNA. As a result, the hormone activates the transcription of the DNA to form mRNA, which in turn translates to synthesize specific proteins in the cytoplasm to perform the comprehensive cellular response (figure 1.15).

1.17 RECEPTOR, HORMONE AND ANTAGONIST

As stated earlier, the signal molecule is called a *ligand* or *agonist* that has specific relationship with its receptor. It is a well-known fact that most of the receptors are protein molecules to which the ligand binds and further activates the enzyme adenylate cyclase that in turn induces changes in the cellular system for response via cAMP. Agonists are the hormones that activate receptors, while *antagonists* have no activating effect on the receptors. When the cells are continuously exposed to agonists, there can be a temporary reduction in the number of receptors and the cells can be desensitized. As a result cells respond less strongly to agonists. The phenomenon is called *down regulation*. In addition, phosphorylation can inactivate the receptor on its cytoplasmic side so that, when the agonist binds with its specific receptor, the production of cAMP can be hampered. Since this phoshorylation is itself cAMP-stimulated, the system forms a logical cAMP feedback inhibition of its own production. There are several hormone receptors like those for epinephrine, insulin and glucagon often undergo down regulation and as a result, the cells show diminished response.

Antagonists are the blockers for receptors and they are sufficiently similar to natural agonists to be 'recognised' by the receptor and to occupy without activating it, thereby preventing the natural agonist from exerting its effect. Prolonged contact with an antagonist leads to formation of new receptors. This phenomenon is called *up regulation*, which can be explained for the worsening of angina pectoris (the painful condition when the coronary arteries fail to deliver enough blood and oxygen to the cardiac muscles) or cardiac ventricular dysrhythmia (irregular heart beat) in some patients from whom a β-adrenoreceptor blocker has been withdrawn abruptly and the increased level of circulating catecholamines now have access to an increased population of β-adrenoreceptors.

SYNOPSIS

❍ Hormones are chemical messengers secreted by endocrine glands and are transported via bloodstream to their target organs. Some of the secretions diffuse only short distance and act on local cells (paracrine secretion), while some are self-stimulatory, that is, they bind to the receptors on the same cell that is releasing it (autocrine secretion).

❍ Hormones induce physiological functions in the cells thereby helping the organism to adjust to the new situation.

❍ Chemically hormones are classified into three groups:

    ❍ Steroid hormones are lipid-soluble and derived from cholesterol

    ❍ Peptides or proteins

    ❍ Amines derived from amino acids.

❍ Once the function of hormone is over, they are readily destroyed by enzymatic action, inactivated or excreted.

○ Hormones are not species-specific because hormones extracted from animals are found to be effective in man.

○ Cell signalling involves detection of the stimulus (in the form of ligand or agonist) by a specific receptor present within the cell membrane or by the intracellular receptor and triggering the cascade of reactions to execute specific cellular response.

○ Steroid hormones and thyroxine, being lipid-soluble, are transferred through lipid bilayer of the cell without difficulty and their receptors are intracellular while water-soluble hormones cannot pass through lipid bilayer hence they bind to membrane receptors.

○ Signal transduction is a complex process involving transfer of information along its pathways having the machinery for physically transporting substances from one side of membrane to the other with the help of a chain of distinct proteins present in the plasma membrane.

○ Signal transduction pathways have the enzymes, protein kinases and protein phosphatases, whose catalytic actions change the conformations of the next protein in the series further activating or inhibiting the protein to show specific response.

○ Most of the hormones (acting as first messengers) use cAMP as second messenger, a substance that is released into the interior of the cell as a result of the binding of a first messenger—a hormone or other ligand—to a receptor at the outer surface of the cell.

○ Cyclic adenosine monophosphate (cAMP) is produced from ATP by an enzyme, adenylate cyclase that is attached to the inside of the cell membrane.

○ Cyclic AMP is produced in the cytoplasm and in turn activates the PKA (protein kinase A). Some of the activated PKA play a significant role in activation of genes by phosphorylating nuclear proteins called CREB (cAMP-responsive element binding protein). The phosphorylated CREB becomes an active transcriptional factor that binds to the DNA having a particular sequence of nucleotides that are known as the cAMP-regulated enhancer (CRE).

❍ The effect of hormone receptor complexes on adenylate cyclase is indirect. It is mediated by a third membrane protein called GTP-binding protein, or trimeric G protein that is associated with the cytoplasmic face of the receptor protein having three polypeptides a, b, and g.

❍ Cyclic GMP is analogous to cAMP that can be used by a diverse group of hormones as second messenger. The cGMP is formed from GTP by the action of guanylate cyclase enzyme. It mediates the cell response by activating specific protein kinases. The effects produced by cGMP are more specialized than those of cAMP. Relaxation of smooth muscles in blood vessels of heart, effects on nerve cells and vision are mediated by cGMP.

❍ Calcium ions play an important role in varieties of physiological activities like cellular movement, endocytosis, metabolism, secretion, fertilization, synaptic transmission, and cell division. It does not act on different targets in free ionic state, but instead its role as a second messenger is associated with a widely distributed protein known as calmodulin. Depending on the type of cell, the Ca^{2+}—calmodulin complex may activate protein kinase, a cyclic nucleotide phosphodiesterase, or even to calcium-transport system of the cell membrane subsequently to show their multiple cellular responses.

❍ The liver and muscle cells show response to insulin due to the presence of insulin receptors in their plasma membrane. The insulin receptor is also an enzyme, a protein tyrosine kinase, that phosphorylates the tyrosine residue present in the other protein, which further phosphorylates a variety of insulin receptor substrates (IRSs). The phoshorylation of the IRS proteins in turn activate the downstream signalling pathways like a) stimulation of the movement of glucose transporters to the plasma membrane to import additional glucose molecules, b) activation of the transcriptional factors for expressing insulin-specific genes, c) promoting DNA synthesis and further cell division, or d) activating genes for protein synthesis.

❑ Phosphotidylinisitol-derived second messengers in the form of inisitol 1,4,5-triphosphate (IP_3) and diacylglycerol (DAG) form one of the signal transduction pathways, which may be activated by G protein-coupled receptors.

❑ The receptor for glucocorticoid is located in the cytoplasm while sex steroid receptors are present in the nucleus. Most of the steroid hormones increase the growth, development and differentiation of tissues that are associated with increase in the rate of protein synthesis or the formation of the enzymes for performing specific cellular function. Steroid hormones have a regulatory impact on the genes through transcriptional factors.

❑ Antagonists are the blockers for receptors which thereby prevent the natural agonist from exerting its effect. Prolonged contact with an antagonist leads to formation of new receptors.

REVIEW QUESTIONS

1. Define hormone and give the difference between exocrine and endocrine secretion.

2. Explain how paracrine secretion differs from autocrine secretion.

3. Enumerate different classes of hormones with their properties.

4. Explain different modes of action of hormones for triggering cellular responses.

5. What is cell signalling? Describe the role of various receptors in cell signalling.

6. How does cAMP exert its effect as second messenger?

7. Explain signal transduction pathways for water-soluble and lipid-soluble hormones.

8. Describe the process of activation of genes through protein kinases.

9. What is the role of G protein in mediating the role of cAMP?

10. What is Ca^{2+}–calmodulin system? Explain its role in execution of cellular responses.

11. Define receptor tyrosine kinases. How do they involve in response of muscle and liver cells to insulin?

12. Explain the mode of action of steroid hormones.

13. Write short notes on

    i.   Nitric acid as second messenger

    ii. Growth factors

    iii. Catecholamines

    iv. Phospolipase C

    v. Cholera and GTP hydrolysis

TWO

HYPOTHALAMUS AND PITUITARY GLAND

2.1 STRUCTURE OF HYPOTHALAMUS AND PITUITARY GLAND

For the well-being of an organism, there are two major administrator units in the body—the nervous system and endocrine system—and both the systems are interrelated functionally. Almost all vertebrates have endocrine glands or organs which play an indispensable role in many of the physiological activities including metabolism, osmoregulation, growth, reproduction, etc. The pituitary gland or hypophysis forms the major endocrine organ in most vertebrates. There are certain modified motor neurons acting for neurosecretion in the hypothalamus—a part of the forebrain forming the functional link between the endocrine control and nervous coordination of the body.

The vertebrate forebrain has two major divisions—*cerebrum* and *diencephalon*. The latter consists of dorsal *thalamus* and ventral *hypothalamus*. The *hypothalamus* is mainly concerned with regulation of body temperature, water balance, feeding, drinking, emotion and reproductive behaviour through autonomic nervous control. Stimulation of hypothalamus can elicit thirst, hunger, rage, sexual drive, pleasure or pain. In addition, it is also a seat of neurosecretions, as it secretes a series of neurosecretory substances that have regulatory influence over the endocrine function of the pituitary gland.

The *pituitary gland* or *hypophysis* is a reddish-grey coloured compact mass of cells having the size of the tip of the little finger. It is situated in the hypophyseal fossa (sella turcica) of the sphenoid bone at the base of the brain. It is divided into two completely separate parts, the *anterior pituitary gland* or *adenohypophysis* and *posterior pituitary gland* or *neurohypophysis* that is connected by a stalk with the hypothalamus of the brain.

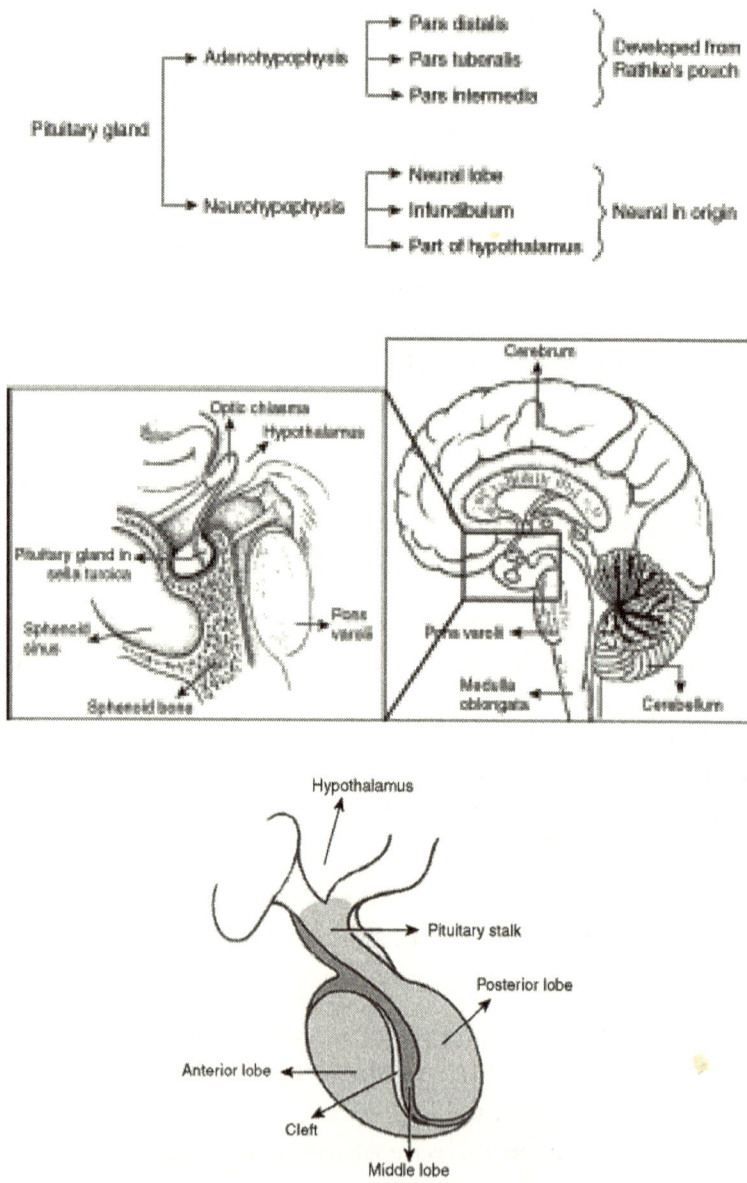

Figure 2.1   Structure of pituitary gland and associated parts in human brain.

Both the subdivisions are distinctly different in embryonic origin and in histological composition. From the embryonic diencephalon, there extends downward a hollow, fingerlike process called infundibulum. From the embryonic mouth, there grows upward an ectodermal pocket called hypophyseal pouch (Rathke's pouch). Both these embryonic structures proliferate masses of the tissues and unite to form the adult pituitary or hypophysis. The adenohypophysis is derived from the hypophyseal pouch while the neurohypophysis is formed from brain tissue. The adenohypophysis includes the anterior lobe (*pars distalis*), the major portion of the whole gland, with its *pars tuberalis,* and the intermediate lobe (*pars intermedia*). The neurohypophysis consists of posterior lobe or neural lobe that also includes certain nuclei in the hypothalamus. The axons of these nuclei descend into the neural lobe through the stalk, the infundibulum. Thus, a fraction of the hypothalamus forms an integral part of the posterior lobe of the pituitary gland (figure 2.1).

HORMONES OF PITUITARY GLAND

As stated earlier, the pituitary gland has two parts that are different from each other in their origin and function—the adenohypophysis and neurohypophysis.

2.2 ADENOHYPOPHYSIS OR ANTERIOR LOBE OF PITUITARY GLAND

The anterior pituitary gland or adenohypophysis is composed of different types of epithelial cells that can be distinguished microscopically by their size and shape. Adenohypophysis secretes at least six different hormones, which stimulate the production of hormones by other endocrine glands or the growth of the body as a whole (figure 2.2). These six hormones are the following:

1. Growth hormone (GH) or somatotrophic hormone (STH)

2. Thyrotropin or thyroid stimulating hormone (TSH)

3. Adrenocorticotrophic hormone (ACTH)

4. Prolactin or lactogenic hormone or luteotrophic hormone (LTH)

5. Follicle stimulating hormone (FSH)

6. Luteinizing hormone (LH) or interstitial cells stimulating hormone (ICSH)

The last two hormones are called gonadotropic hormones since they regulate the functions of the testes and ovaries.

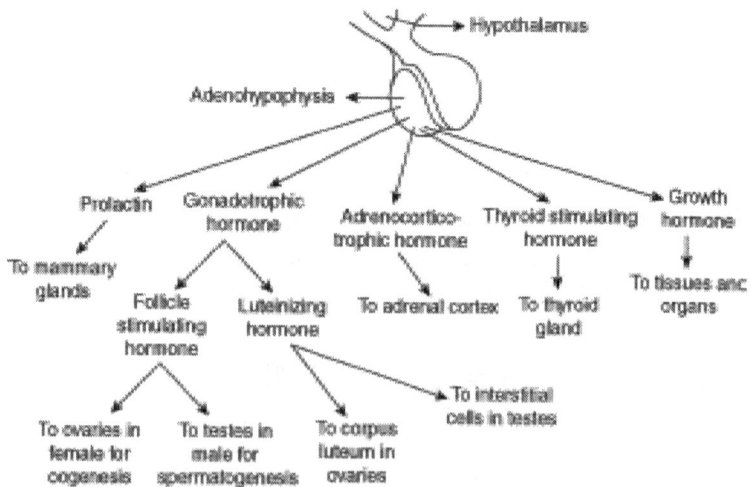

Figure 2.2   Hormones of anterior pituitary gland (adenohypophysis) with their target organs.

2.2.1 Growth Hormone (GH) or Somatotrophic Hormone (STH)

GH or STH is a small peptide hormone having 191 amino acids in its polypeptide chain. In man, its molecular weight is 22,500 while different groups of animals have its molecular weight ranging from 21,500 to 48,000. The production of growth hormone does not stop even though most of the growth in the body ceases at adolescence.

Functions of growth hormone

1. GH promotes the development and enlargement of all tissues of the body.

2. It stimulates the growth of long bones by promoting multiplication of the epiphyseal cartilage.

3. GH enhances protein anabolism in the body. The increased rate of protein synthesis strongly influences the growth of the muscles, bones and visceral organs. Under its impact, intestinal absorption of calcium, conversion of glycogen to glucose, release of fatty acids from adipose tissue, secretion of milk during lactation, etc. are promoted.

4. The continuous supply of growth hormone is essential for life, to stimulate the repair and replacement of body tissue.

Disorders   The normal growth of an individual is controlled considerably by STH; hence its amount in the blood and rate of secretion varies from age to age. If the secretion of this hormone is decreased (hyposecretion) during early life, it leads into cessation of growth and results in *dwarfism*. On the other hand, increased secretion of GH (hypersecretion) during early life produces abnormal growth of bones as a result there is abnormal increase in height or length of the body, the condition is called *gigantism*. In adults excessive production of STH leads to *acromegaly*.

Dwarfism   It is the abnormal condition that occurs in children due to hyposecretion of GH. It is characterized by delayed skeletal growth and retarded sexual development but they are alert, intelligent and having well-proportionate body (figure 2.3). Increased adiposity in a child who is growing poorly suggests the possibility of growth hormone deficiency or hypothyroidism.

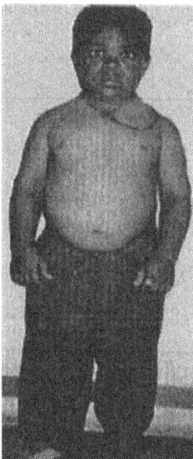

Figure 2.3 Dwarfism resulted due to hyposecretion of growth hormone.

Figure 2.4 Basket ball player in U.S. with height 7′ 7″. His gigantism is due to hypersecretion of GH.

Gigantism   Hypersecretion of GH during childhood results in this rare disorder that is characterized by overgrowth of all body tissues and organs particularly bones. The individual with this disorder may attain the height as much as 2.6 metres

(8.5 feet). Such abnormal increase in height is due to excessive growth of limb bones (figure 2.4). They are often mentally retarded and may die before the age of 20 due to lack of proper treatment.

Acromegaly The condition arises in adults due to excessive secretion of STH. The patient has gorilla-like facial appearance with increased fullness and coarsening of soft tissues, including lips and tongue and also thickening of skin and sub-cutaneous tissues and progressive enlargement of head, hands and feet [figures 2.5 (a) and (b)]. In addition, the hands of the patient are massive, the fingers spatulate with square tips and skin thickened.

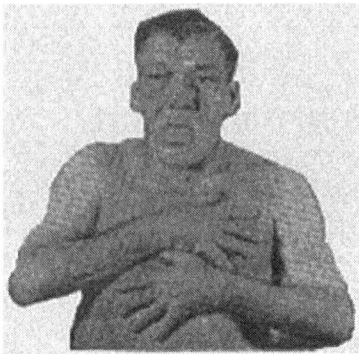 

(a)   **(b)**

Figure 2.5 Two patients of acromegaly showing enlargement of lower jaws, hands (a), thickening of nose and ridges above the eyes (b) due to oversecretion of growth hormone in adult.

2.2.2 Thyrotropin or Thyroid Stimulating Hormone (TSH)

TSH is glycoprotein in nature with a molecular weight of about 25,000. It is secreted by the basophils of adenohypophysis.

Functions   TSH controls various functional aspects of the thyroid gland including its growth and maintenance. It influences the uptake of iodine and the synthesis of the hormones *thyroxine* and *triiodothyronine* by thyroid gland and release of stored hormones into the bloodstream.

The secretion of TSH by adenohypophysis and the amount of thyroxine circulating in the blood exerts the negative feedback mechanism. Increased secretion of thyroxine from thyroid gland due to thyrotropin secondarily leads to increase in basal metabolic rate.

Disorders   If the secretion of TSH is decreased due to some disease of adenohypophysis, thyroid gland will also decrease its secretion as a result the animal suffers from *hypothyroidism*. Such disorder produces symptoms similar to those of primary hypothyroidism. The condition is called *myxoedema*. On the other hand, if TSH from anterior pituitary is increased, the activity of thyroid gland also increases, producing the condition called *hyperthyroidism*. This condition is clinically characterised by bulging of eyeballs or *exophthalmos*. Both types of disorders are later discussed in detail.

2.2.3 Adrenocorticotrophic Hormone (ACTH)

ACTH or corticotropin is protein in nature having 39 amino acids in its polypeptide chain with a molecular weight of about 4,500. This hormone has the following functions:

1. It stimulates the cortex of the adrenal gland to produce the hormones—*mineralocorticoids, glucocorticoids* and *sex steroids*.

2. It also influences protein synthesis in adrenal cortex.

3. It has lipolytic effect on adipose tissues.

4. Corticotropin is essential for maintaining the normal activities of adrenal cortex.

5. ACTH and hormones of the adrenal cortex (specifically glucocorticoids) have a significant role to play in preparing an animal to meet stressful stimuli such as pain, injury, excess heat or cold, haemorrhage, etc.

6. It has an intrinsic melanocyte stimulating activity since out of 39 amino acids in its composition, a series of 13 amino acids are common to that of MSH.

Disorders   Pituitary dysfunction may cause hypersecretion of ACTH that in turn induces hyperplasia (cell multiplication) and hyperactivity of adrenal cortex. This rare clinical condition is called *Cushing's disease*, which occurs mainly in females. It is characterised by *virilism* (appearance of secondary male features in female), obesity, hyperglycemia, glycosurea and hypertension.

2.2.4 Prolactin or Lactogenic Hormone or Luteotrophic Hormone (LTH)

Prolactin is a hormone secreted by the anterior pituitary gland during the last few days of pregnancy and during the entire period of milk production (lactation) in women after the birth of the baby. LTH is a peptide hormone having a molecular weight of about 25,000. It has the following functions:

1. Lactogenic hormone stimulates production of milk in the mammary gland (Although the breasts are fully developed, lactation does not occur during pregnancy. It starts only after parturition and expulsion of placenta. The release of milk from breast is primed by several hormones).

2. LTH influences the development of the corpus luteum in the ovary.

3. It stimulates crop gland in birds like pigeon and doves to produce crop milk.

4. Prolactin has an influencing role in changing the maternal behaviour that is essential for the survival of helpless new-borns.

5. In some birds, LTH promotes nesting behaviour in both sexes.

6. In certain animals, prolactin is responsible for migration where they breed or become sexually mature.

Thus, prolactin is an essential hormone for normal breeding behaviour, pregnancy and lactation. Its role in males is obscure.

2.2.5 Gonadotrophic hormones (GTH) or Gonadotropins

The two gonadotropic hormones namely follicle stimulating hormone (FSH) and luteinizing hormone (LH) are the hormones that are essential for development, maintenance and hormonal function of the gonads—testes and ovaries.

2.2.6 Follicle Stimulating Hormone (FSH)

It is a water-soluble glycoprotein whose molecular weight ranges from 30,000 to 67,000. In the female, FSH initiates growth of follicle in the ovaries. The development and maturation of ovarian follicle until its ovulation is under the influence of FSH. It also helps the ovaries to secrete *oestrogens*, one of the female sex hormones.

In the male, follicle stimulating hormone stimulates the germinal epithelium in the testes to promote the development of sperm.

2.2.7 Luteinizing Hormone (LH) or Interstitial Cell Stimulating Hormone (ICSH)

It is a conjugate protein with a molecular weight ranging from 26,000 to 30,000.

In the female, the peak level of LH in blood causes ovulation i.e. rupture of mature ovarian follicle to release ovum into the body cavity and then into the fallopian tube. It is also responsible for the development of corpus luteum, which in turn secretes *progesterone*, the hormone essential for maintenance of pregnancy.

In the male, ICSH promotes the interstitial cells or Leydig's cells in testes to secrete testosterone, the male sex hormone that plays a key role in the maintenance of accessory reproductive organs and secondary sexual characters.

2.3 PARS INTERMEDIA OR MIDDLE LOBE OF PITUITARY GLAND

It is the smallest and the middle part of the pituitary gland that consists of a thin strip of epithelial tissue and is separated from the anterior pituitary by the interglandular cleft. It is well developed in poikilotherms (fishes, amphibians and reptiles) while in homeotherms it seems to have little functional importance.

Pars intermedia secretes two types of *melanocyte stimulating hormones* (MSH) namely α-MSH and β-MSH; the α-MSH has

13 amino acid residues while β-MSH has 18 amino acids. They affect pigment dispersion in the melanophores of lower vertebrates, thus being responsible for producing characteristic colour patterns. When the pituitary is removed from the frog, it results in permanent blanching of the skin; this is because the melanin is concentrated within melanophores and MSH is essential for expansion of melanophores. Thus, MSH initiates the changes that are necessary for darkening of the skin.

2.4 NEUROHYPOPHYSIS OR POSTERIOR LOBE OF PITUITARY GLAND

Physiologically, the posterior lobe of the pituitary gland is commonly called the pars nervosa forming the principal part of the neurohypophysis. It is a neurohaemal organ composed

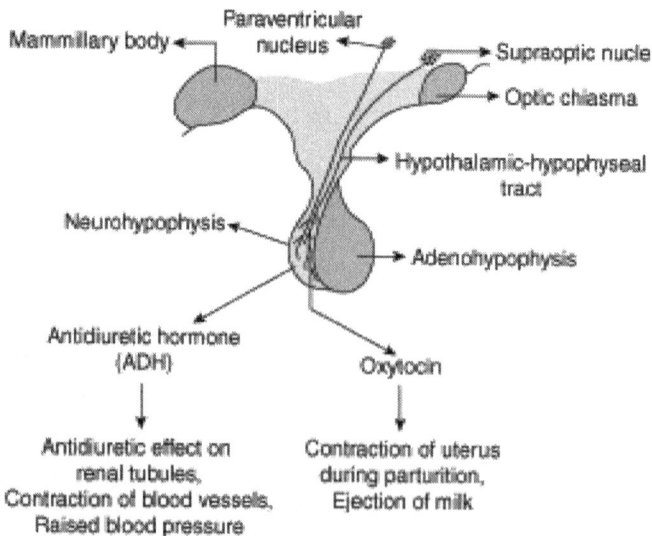

Figure 2.6 Hormones released by neurohypophysis with their functions.

of secretory cells called *pituicytes* and nerve fibres that arise from cells in two areas of the hypothalamus called *supraoptic nuclei* and *paraventricular nucleus*. Neurohypophysis and hypothalamus form a single functional unit and control various functions in the body by secreting two important hormones—antidiuretic hormone (ADH) and oxytocin. Both hormones are secreted by supraoptic and paraventricular nuclei in the hypothalamus and they migrate along the tracts of the nerve fibres up to the posterior pituitary where these hormones are stored in the nerve endings (figure 2.6). Each hormone is released in response to a specific stimulus.

2.4.1 Antidiuretic Hormone (ADH)

It is also called vasopressin and is formed primarily in the supraoptic nuclei of hypothalamus. It is an octapeptide hormone with a molecular weight of 1100 and containing eight amino acids in its structure. It has the following important biological functions:

Antidiuretic effect It increases the reabsorption of water in the kidney tubules. It mainly acts on distal convoluted tubules and renal collecting tubules of the nephrons of the kidneys and increases their permeability to water. As a result the reabsorption of water from glomerular filtrate is increased. The secretion of ADH is regulated by the osmoregulatory centre in the hypothalamus. When the osmotic pressure of the blood rises, the secretion of ADH increases and more water is reabsorbed. Conversely, when there is reduction in osmotic pressure of the blood, less amount of ADH is released and subsequently less water is reabsorbed and more diluted urine is produced.

Pressor effect   Under the influence of ADH, involuntary muscles present particularly in arteriovenous capillaries undergo contraction. As a result blood pressure increases and hence ADH is also called vasopressin. And this is called pressor effect.

Disorder due to lack of ADH   When ADH is not secreted or it is secreted in reduced amount, water reabsorption from kidney tubules is decreased leading to formation of dilute urine. The patient excretes large volumes of diluted urine (polyuria) and to compensate this water loss he frequently drinks more water (polydipsia). These are the symptoms of the physiological disorder called *diabetes insipidus*. The condition can be corrected by administration of vasopressin.

2.4.2 Oxytocin or Pitocin

It is also an octapeptide hormone with a molecular weight of about 1000. It is primarily synthesized in the paraventricular nucleus of the hypothalamus. It is so called because it produces oxytocic effect on the pregnant uterus so the term oxytocin is derived from the Greek word, meaning 'quick birth'.

Effect of oxytocin on uterus   An oxytocic agent is a substance that causes the uterus to contract. This is one of the primary functions of oxytocin. It is secreted in moderate quantities during the latter part of pregnancy and in especially large quantities at the time of delivering the baby. It probably helps in the expulsion of the baby from the uterus. It is well known that removal of pituitary gland from pregnant female results in considerable difficulty and protracted parturition.

Effect of oxytocin on mammary glands   Oxytocin encourages the process of lactation, for it brings about ejection of milk by initiating the contraction of smooth muscles in mammary glands and myoepithelial cells that are present around the alveoli and induces expulsion of milk. This action is called *galactogogue* effect. When oxytocin is injected in a lactating cow, it starts gushing milk from its udder. Thus oxytocin plays an indispensable role in milk ejection.

In addition to initiation of uterine contraction during parturition, sexually stimulated females during coitus secrete small amounts of oxytocin that induce uterine contraction. This helps in the ascent of sperms in the fallopian tubes and thus facilitates fertilization.

2.5 HYPOTHALAMIC REGULATION FOR RELEASE OF PITUITARY HORMONES

As stated earlier, physiologically pituitary gland has two parts: adenohypophysis or anterior lobe and neurohypophysis or posterior lobe. The adenohypophysis is a highly vascular organ and secretes six different hormones that play a major role in control of various metabolic functions in the body. Adenohypophysis receives blood supply from two sources: (a) the usual arteriolar supply and (b) the *hypothalamic-hypophyseal portal system*. The hypothalamus is the part of the fore brain lying immediately below the thalamus and controls many of the automatic functions of the body. After the blood passes through of the hypothalamus, particularly through the lower part known as *median eminence*, it leaves via the small *hypothalamic-hypophyseal portal veins* that course down the anterior surface of the pituitary stalk to the adenohypophysis. At this point, the veins dip into the gland where the blood flows through large numbers of *venous sinuses* that bathe the adenohypophysis (figure 2.7).

The hypothalamus secretes a series of different neurosecretory substances known as hypothalamic releasing and inhibitory factors or simply neurohormones. These factors are secreted into the blood of the hypothalamic-hypophyseal portal system and are then transported to venous sinuses in the anterior pituitary gland where they regulate secretory activity of the adenohypophysis. There are six prominent factors secreted by the hypothalamus that control the anterior pituitary secretion. These hypothalamic factors are as follows:

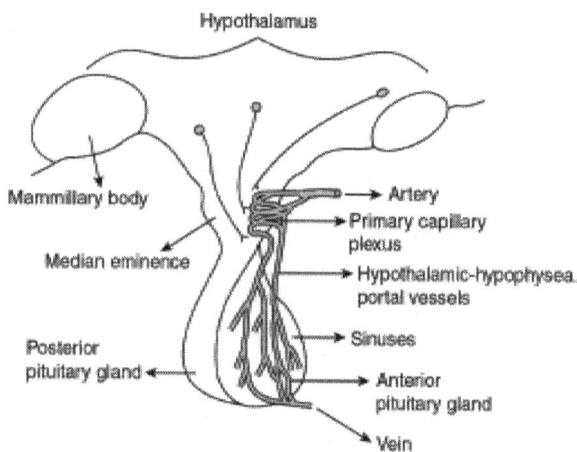

Figure 2.7 The hypothalamic-hypophyseal portal system.

1. *Thyrotropin-releasing factor (TRF)*   It stimulates anterior pituitary to release thyroid stimulating hormone.

2. *Corticotropin-releasing factor (CRF)*   It causes release of adrenocorticotrophic hormone or corticotropin from anterior pituitary.

3. *Growth hormone-releasing factor (GRF)*   It encourages adenohypophysis to release growth hormone.

4. *Luteinizing hormone-releasing factor (LRF)*   It stimulates anterior pituitary to secrete luteinizing hormone.

5. *Follicle stimulating hormone-releasing factor (FRF)* It causes release of follicle-stimulating hormone from adenohypophysis.

6. *Prolactin inhibitory factor (PIF)* It causes inhibition of prolactin secretion from anterior pituitary gland.

It is important to note that out of six factors mentioned above, five factors stimulate the anterior pituitary to secrete its respective hormones. In the absence of TRF, CRF, GRF, LRF and FRF, pituitary cannot secrete its respective hormones in large quantities. On the other hand, PIF is an inhibitory factor. In the absence of prolactin inhibitory factor, adenohypophysis secretes excess of prolactin.

Thus the secretion of hormones by the anterior pituitary is largely under the control of neurohormones from the hypothalamus. Because of the powerful effect of these factors on the adenohypophysis, major control of the anterior pituitary gland is exercised through stimulation or inhibition of the many neuronal control centres in the hypothalamus that in turn regulate the output of the hypothalamic releasing or inhibitory factors. For instance, thyrotropin-releasing factor (TRF) from hypothalamus encourages anterior pituitary to secrete thyroid stimulating hormone (TSH) that in turn stimulates thyroid gland to secrete thyroxine and triiodothyronin, provided the hypothalamus is stimulated to do so. The increased output of thyroxine accelerates tissue metabolism which results in an increase in body temperature that inhibits hypothalamic centres and thus controls the secretion of TRF. The control of many of the other hormones is affected through a similar hypothalamic-hypophyseal system that we shall see in succeeding chapters.

Another evidence to prove the influencing role of hypothalamus over the secretory activity of anterior pituitary, comes from the grafting experiments. The grafting of female pituitary under the male hypothalamus leads to secretion of the male gonadotropins, while the grafting of male adenohypophysis under female hypothalamus results in the release of female gonadotropins indicating the dominating role of hypothalamus in controlling secretory activity of hypophysis. The hypothalamus receives information about conditions in the body and in the external environment through both neuronal and hormonal signals. If the connection between the hypothalamus and the pituitary is experimentally cut, pituitary hormones are no longer released in response to changes in the internal or external environment. When pituitary cells were maintained in culture, extract of hypothalamic tissue stimulated some of the cells of pituitary to release their hormones in the culture medium.

In relation to the secretions of the posterior pituitary gland, it has been already mentioned that though the neurohypophysis is located immediately behind the adenohypophysis, both glands do not have any proved direct relationship. Also, in a sense, the posterior pituitary gland is not a gland at all, because it only stores hormones rather than synthesizing them. Both antidiuretic hormone and oxytocin are secreted by neuronal cells in supraoptic nucleus and paraventricular nucleus respectively that are present in the anterior part of hypothalamus and then are conducted through nerve axons to the posterior pituitary gland where both hormones are stored. The anatomical features for this mechanism are shown in the figure 2.6.

Thus, though the pituitary gland is considered as the *master endocrine gland* as far as the secretion of various hormones is concerned, its secretory activity is altogether controlled by the hypothalamus.

SYNOPSIS

❍ The *hypothalamus* is a part of forebrain that is mainly concerned with regulation of body temperature, water balance, feeding, drinking, emotion and reproductive behaviour through autonomic nervous control.

❍ The *pituitary gland* or *hypophysis* is situated in the sella turcica of the sphenoid bone at the base of the brain. It is divided into the *anterior pituitary gland* or *adenohypophysis* and *posterior pituitary gland* or *neurohypophysis* that is connected by a stalk with the hypothalamus of the brain.

❍ Adenohypophysis secretes six different hormones namely growth hormone (GH), thyroid stimulating hormone (TSH), adrenocorticotrophic hormone (ACTH), prolactin (PL), and luteinizing hormone (LH) or interstitial cells stimulating hormone (ICSH) that stimulate the production of hormones by other endocrine glands or the growth of the body as a whole.

❍ The normal growth of an individual is controlled considerably by GH. Its hyposecretion during early life leads into cessation of growth and results in *dwarfism*. On the other hand, increased secretion of GH (hypersecretion) during early life produces abnormal growth of bones. As a result, there is abnormal increase in height or length of the body, the condition called *gigantism*. In adults, excessive production of GH leads to *acromegaly*.

❍ TSH controls various functional aspects of the thyroid gland including its growth and maintenance. The secretion of TSH by adenohypophysis and amount of thyroxine circulating in the blood exerts the negative feedback mechanism. If the secretion of TSH is decreased due to some disease of adenohypophysis, thyroid gland will also decrease its secretion as a result the animal suffers from *hypothyroidism*. Such disorder is called *myxoedema*. On the other hand, if TSH from anterior pituitary is increased, the activity of thyroid gland also increases, producing the condition called *hyperthyroidism*. This condition is clinically characterized by bulging of eyeballs or *exophthalmos*.

- ○ ACTH stimulates the cortex of the adrenal gland to produce the hormones—*mineralocoticoids, glucocorticoids and sex steroids*. Pituitary dysfunction may cause hypersecretion of ACTH that in turn induces *Cushing's disease*, which occurs mainly in females. It is characterized by virilism, obesity, hyperglycemia, glycosuria and hypertension.

- ○ Prolactin or lactogenic hormone stimulates production of milk in the mammary gland and influences the development of the corpus luteum in the ovary.

- ○ Gonadotropins are the hormones that are essential for development, maintenance and hormonal function of the gonads—testes and ovaries. Follicle-stimulating hormone (FSH): initiates growth of follicle in the ovaries in female and helps the ovaries to secrete *oestrogens*. In male, it stimulates the germinal epithelium in the testes to promote the development of sperm. In female, the peak level of luteinizing hormone (LH) in blood causes ovulation and stimulates development of corpus luteum, which in turn secretes *progesterone*—the hormone essential for maintenance of pregnancy. In male, interstitial cells stimulating hormone (ICSH) promotes the interstitial cells or Leydig's cells in testes to secrete testosterone, the male sex hormone.

- ○ Pars intermedia or middle lobe of pituitary gland is well developed in poikilotherms while in homeotherms it seems to have little functional importance. It secretes two types of *melanocyte stimulating hormones* (MSH) α-MSH and β-MSH that affect pigment dispersion in the melanophores of lower vertebrates, and are thus responsible for producing characteristic colour patterns.

- ○ Neurohypophysis or posterior lobe of pituitary gland is a neurohaemal organ that stores two hormones, which are synthesized in *supraoptic nuclei* and *paraventricular nucleus* of hypothalamus. These two hormones are antidiuretic hormone (ADH) and oxytocin.

❑ Antidiuretic hormone (ADH) is also called vasopressin. It shows antidiuretic effect by increasing reabsorption of water in the kidney tubules. The secretion of ADH is regulated by osmoregulatory centre in hypothalamus. It also causes contraction of blood vessels thereby increasing blood pressure. Absence of ADH leads to the physiological disorder called *diabetes insipidus* that is characterized by polydipsia and polyurea.

❑ Oxytocin or pitocin helps in parturition by facilitating expulsion of the baby from the uterus. It also stimulates lactation by initiating the contraction of smooth muscles in mammary glands. In addition, it initiates uterine contraction during coitus for ascent of sperms in fallopian tubes thus helping for fertilization.

❑ The hypothalamus secretes a series of different neurosecretory substances known as hypothalamic releasing and inhibitory factors into the blood of the hypothalamic-hypophyseal portal system and that are then transported to venous sinuses in the anterior pituitary gland where they regulate secretory activity of the adenohypophysis. Six prominent factors secreted by hypothalamus that control the anterior pituitary secretion are thyrotropin-releasing factor (TRF), corticotropin-releasing factor (CRF), growth hormone-releasing factor (GRF), luteinizing hormone-releasing factor (LRF), follicle-stimulating hormone-releasing factor (FRF) and prolactin inhibitory factor (PIF). While posterior pituitary or neurohypophysis is not a gland to synthesize any hormones, it stores the hormones that are synthesized in the hypothalamus.

REVIEW QUESTIONS

1. Explain the structural integrity of hypothalamus and pituitary gland.

2. Describe the role of hormones secreted by adenohypophysis.

3. What is dwarfism and gigantism?

4. Explain the features of the disorder, acromegaly.

5. Describe the role of follicle stimulating hormone, luteinizing hormone and interstitial cell stimulating hormone in male and female.

6. What is antidiuretic effect? Explain the disorder caused by lack of antidiuretic hormone.

7. Explain how hypothalamus controls the secretory activity of the pituitary gland.

8. Write short notes on:

   i.   Thyroid stimulating hormone

   ii.   Prolactin inhibitory factor

   iii. Supraoptic and paraventricular nuclei

   iv.   Melanocyte stimulating hormone

   v.   Cushing's disease

   vi.   Hypothalamic-hypophyseal portal system

   vii. Diabetes insipidus

   viii. Oxytocic effect

   ix.   Releasing and inhibitory factors from hypothalamus

THREE

THYROID AND PARATHYROID GLANDS

THYROID GLAND

The thyroid gland is situated in the neck in the form of two lobes on either side of trachea. A narrow band of *isthmus* connects both the lobes (figure 3.1).

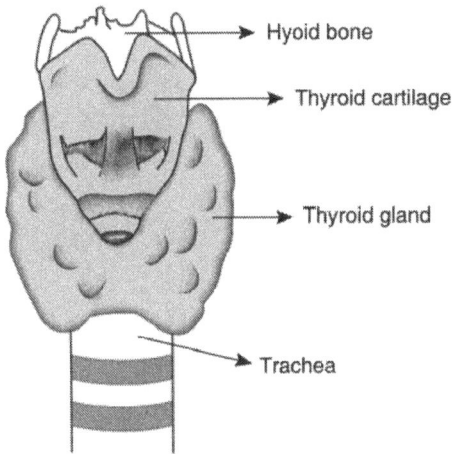

Figure 3.1 Thyroid gland with its associated structures.

3.1 STRUCTURE OF THYROID GLAND

The average weight of human thyroid is about 30 to 40 gm. It is a highly vascular gland since it receives ample supply of blood (about 5 lit/hr) through the branches of external carotid and subclavian arteries. Histologically, the human thyroid is composed of a large number of spherical glandular follicles or vesicles that are lined by cuboidal epithelium (figure 3.2). The thyroid vesicles are filled with a colloidal material called *thyroglobulin*. It is an iodized glycoprotein with a molecular weight of about 680,000 and contains the hormones *thyroxine*

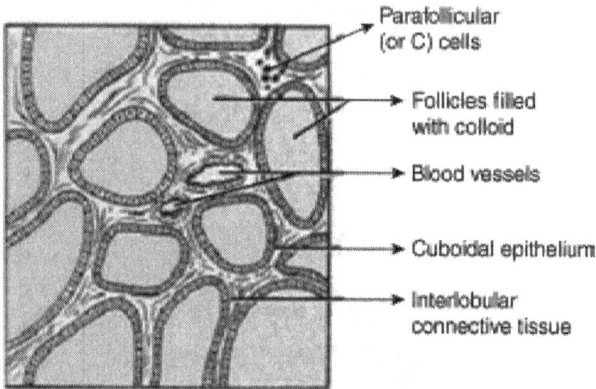

Figure 3.2 Histological features of thyroid gland.

and *triiodothyronine*. The parafollicular cells or 'C' cells present in the interlobular space secrete the hormone called *calcitonin* or *thyrocalcitonin*.

3.2 BIOSYNTHESIS OF THYROID HORMONES

The tissues of thyroid gland have high affinity towards iodine. The food supplemented with iodine (from iodized salt) is digested and absorbed in the gastrointestinal tract from where iodine is transported to the thyroid gland via blood circulation. The follicular cells absorb iodine in the form of ionic iodine (I^-), which is then converted into elemental iodine. Further this elemental iodine combines with the amino acid, *tyrosine* to form *monoiodotyrosine* (MIT). Further the MIT is iodinated to form diiodotyrosine (DIT). Addition of an iodine to DIT leads to formation of *triiodothyronine* (T_3). When two molecules of DIT join with each other, with the loss of the alanine side chain, there is the formation of *tetraiodothyronine* (T_4) that is also called *thyroxine*. Their synthetic pathways are shown in figure 3.3.

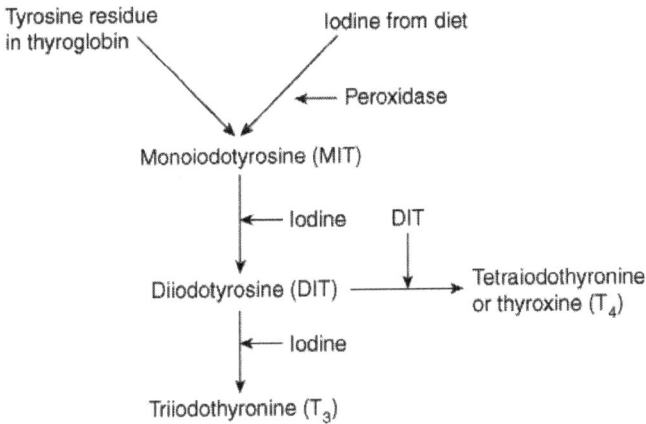

Figure 3.3 Pathways for synthesis of triiodothyronine (T_3) and thyroxine (T_4).

Thyroid hormones thus formed combine with the protein part namely globulin and are stored as thyroglobulin in the form of colloidal material inside the thyroid follicles. When thyroid hormones are released into blood circulation they are carried in a loose complex with plasma protein. This complex dissociates at the site of target tissues and cells where they perform many biological functions. Under usual circumstances, T_4 secretion predominates.

3.3 BIOLOGICAL FUNCTIONS OF THYROXINE

Thyroxine has to play several biological functions that vary considerably from species to species. In homeotherms it is mainly concerned with energy production while for poikilotherms it serves as osmoregulatory agent and also controls metamorphosis. The principle actions of thyroid gland are as follows:

1. Thyroid hormones (both T_3 and T_4) influence heat production in the tissues of the body by uncoupling oxidative phosphorylation i.e., increasing oxygen utilization relative to the rate of formation of high-energy phosphate bonds. Thus each mg of T_3 and T_4 raises basal metabolic rate by about 1000 cal and increases the O_2 consumption in most of the tissues. This effect is known as *calorigenic action*.

2. Thyroxine and triiodothyronine are essential for normal physical and mental development and in the metabolism of protein, carbohydrate and fat.

3. They remove calcium and phosphorus from bones leading to osteoporosis.

4. Thyroxine increases nitrogen excretion and volume of urine through kidneys.

5. Thyroid hormones also have influencing role over the activity of the nervous system and voluntary muscles.

6. The rate and depth of respiration increases under the impact of thyroid hormones since there is increase in BMR.

7. Thyroid hormones act directly on the pacemaker of the heart thereby accelerating the heart beat rate.

8. For the normal development and functioning of gonads thyroid hormones are essential.

9. Thyroxine increases the motility of the gastrointestinal tract and promotes copious flow of digestive juices.

10. The normal texture of the skin is maintained due to thyroid hormones.

3.4 REGULATION OF THYROID SECRETION

As mentioned in chapter 2 (section 2.2.2), the secretion of thyroid hormone from thyroid gland is regulated predominantly by thyroid stimulating hormone (TSH) secreted by the anterior pituitary gland. The secretion of TSH is, in turn, regulated by the hypothalamus.

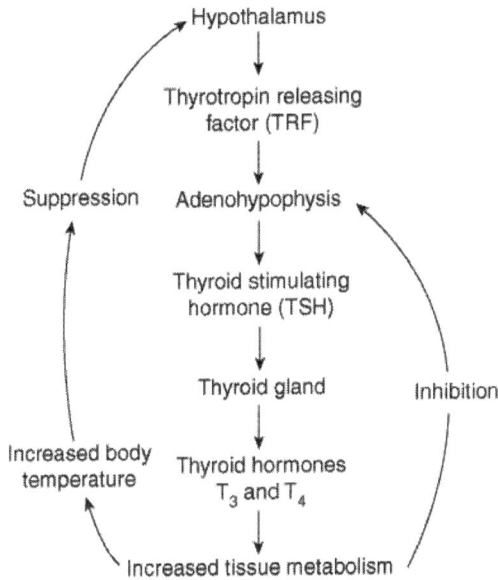

Figure 3.4 Negative feedback mechanism between thyroid gland and adenohypophysis.

The rate of secretion of thyrotropin-releasing factor (TRF) from hypothalamus is primarily controlled by the level of metabolism in the body. If the metabolism falls below normal, the rate of secretion of TRF increases automatically, in turn increasing the secretion of thyrotropin and consequently of thyroid hormones. Thyroxine then accelerates the metabolic rate of the body bringing it to normal. On the other hand, if the rate of metabolism in the body rises above the normal, the hypothalamus decreases the secretion of TRF, and an opposite sequence of events reduces the secretion of thyroid hormones, thereby reducing the metabolic rate to the normal. The interrelationship of pituitary and thyroid is based on the principle of negative feedback mechanism. As the amount of thyroxine rises in the blood, this increased level of thyroxine brings feedback directly on the pituitary to suppress the production of TSH. In the absence of TSH, thyroid gland does not secrete any hormone; as a result there is no further rise in the level of thyroxine (figure 3.4).

3.5 THYROID DYSFUNCTION

When the thyroid gland fails to secrete its normal hormones, the condition is said to be diseased or dysfunction. Excessive secretion of thyroid hormones leads to serious disorder known as *hyperthyroidism* or *thyrotoxicosis* that causes *Grave's disease* or *exophthalmic goitre,* while inadequate secretion results in *hypothyroidism* that has also severe pathological consequences leading to *cretinism* in children and *myxoedema* in adults.

3.5.1 Hypothyroidism

The condition arises due to subnormal secretion of T_3 and T_4 and it depends on the severity of thyroid hormone deficiency and the age of onset of the disorder.

Cretinism   It is an infantile hypothyroidism. Children suffering from this disorder seem normal at birth, having received maternal thyroid hormone while in the uterus, but within few weeks or months, it becomes apparent that mental and physical development are retarded. The milestones of child development, such as holding up the head, sitting, dentition, standing, walking, speech, etc. are delayed. The characteristic cretin is a dwarf with severe mental defect, disproportionately short limbs, coarse dry skin, deficient hair and teeth, a large protruding tongue (figure 3.5), a potbelly and umbilical hernia. Unless replacement therapy with thyroid hormones commences early, the changes, especially the mental defect, become irreversible, though there may be some improvement in physical appearance.

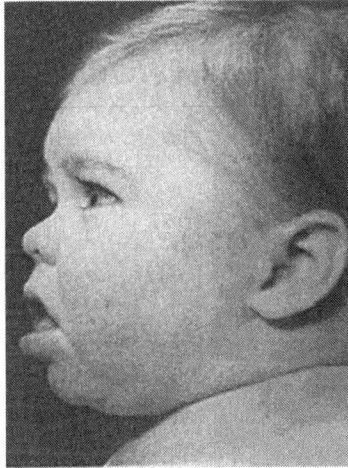

Figure 3.5 A cretin aged 17 months.

Myxoedema Diminished production of thyroxine in adults causes this disorder. The person can live for many years with complete lack of thyroxine production; the rate of metabolism in all his tissues is decreased to about one-half normal. He is extremely lethargic, sleeping sometimes as much as 12 to 15 hours a day. He is constipated and mentally sluggish and often

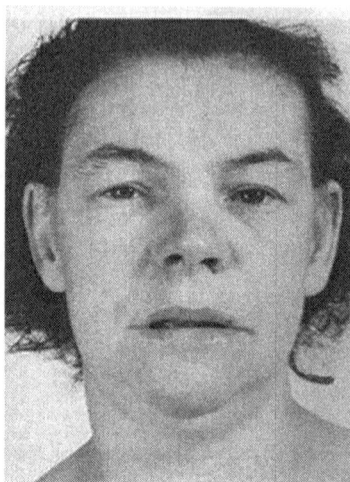

Figure 3.6   Myxoedema—a case of hypothyroidism.

becomes fat despite his poor appetite. The pulse is low and his basal metabolic rate is lowered. In addition, a gelatinous mixture of mucoprotein and extracellular fluid is deposited in the space between the cells, specifically in dermal connective tissues,

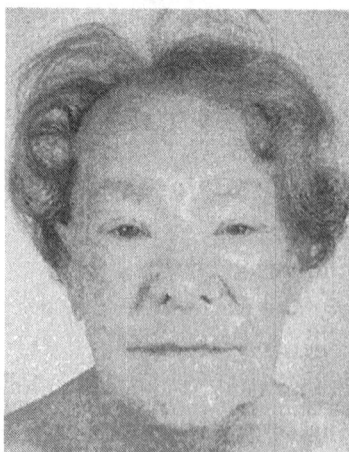

Figure 3.7   Note the facial appearance of hypothyroidism.

giving the oedematous appearance. Hence the condition is often called myxoedema (figure 3.6). Disorder is also associated with abnormal dryness of the skin and the coarseness of the hair (figure 3.7). The hair, brittle and lustreless, tend to fall out. The patients also have slowness in thoughts, speech and actions.

Hashimoto's disease (lymphadenoid goitre)   It is also called Hashimoto's thyroiditis that is an inflammatory condition occurring in 1 to 2 percent of the population. It is the hypothyroidism with diffuse and massive lesion causing goitre that is very common in middle-aged women. The disorder is due to infiltration of thyroid by lymphocytes and plasma cells associated with abnormality of thyroid epithelium. The massive lymphoid infiltration and variable amount of fibrous tissue are very characteristic features of the disease. Hypothyroidism is present in about half the cases, and is the rule if partial thyroidectomy is undertaken. The thyroid uptake of iodine is reduced and organification is impaired.

3.5.2 Hyperthyroidism

The condition is also known as *thyrotoxicosis* that results from excessive secretion of thyroid hormone and causes clinical features which are almost opposite of those found in hypothyroidism. The patient, though weak, is restless and emotionally unstable. Because of higher body temperature and nervousness, he perspires even in winter. The appetite is increased but the patient loses weight. There is increased nitrogen excretion, and sometimes impaired glucose tolerance and glycosuria. The skin becomes soft, warm and moist with increased sweating that helps the loss of heat from the body. The pulse and cardiac output is increased. Osteoporosis results due to excessive loss of calcium and phosphorus from the bones. Many of the above-mentioned features are due to increased basal metabolic rate, that in turn is due to the effect of excessive amount of T_3 and T_4 on the tissues.

One of the most characteristic clinical symptoms of hyperthyroidism, specifically in *Graves' disease,* is the prominence of the eyes (exophthalmos). The patient has a staring look because of protruding eyeballs with less twinkling of the eyelids (figure 3.8), which cannot be attributed to the action of T_3 and T_4.

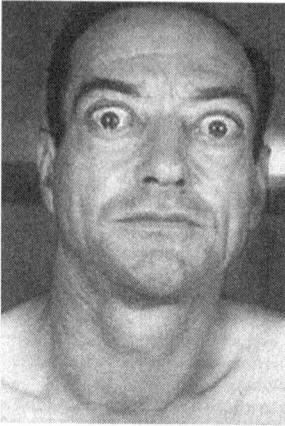

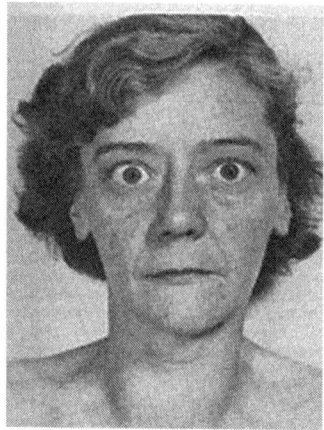

Figure 3.8 Note the prominence of eyes and diffuse thyroid enlargement in thyrotoxic Graves'diesase.

Graves'diesase (exophthalmic goitre) The disorder of thyroid gland caused by excessive secretion of T_3 and T_4, is characterized by diffuse thyroid hyperplasia apparently due to the presence of an inappropriate thyroid stimulator in the blood. Protrusion of eyeballs (exophthalmos) and certain other features as described above cannot be ascribed to thyroid hormones excess alone. Most frequently the increased production of thyroid hormones is caused by a substance called *long-acting thyroid stimulator* (LATS) that is believed to be an antibody formed by the immune system and that has specific reactivity for the thyroid cells. The excess production of thyroxine resulting from this stimulation of the thyroid gland then causes the rate of thyrotropin secretion by the hypothalamus usually to be reduced rather than elevated.

It is now apparent that Graves'diesase is due to an auto-antibody which reacts with the thyroid epithelial cell surface receptor for TSH. This thyroid-stimulating antibody is detectable in the blood of most untreated patients by an *in vitro* test in which the binding of labelled TSH by human thyroid cell membrane preparation is competitively inhibited by the patient's serum.

The patients suffering from hyperthyroidism can be treated with *antithyroid agents* like methyl thiouracil, carbimazole, perchlorate, etc., or by partial surgical thyroidectomy.

Nontoxic goitre (simple goitre)   It is the non-inflammatory condition of the thyroid gland that result in the enlargement of the thyroid gland without hyperthyroidism. There is impaired thyroid hormone synthesis due to inadequate supply of iodide or impaired thyroid enzyme activity, the result of genetic defect or exogenous toxic substances. In India, the tribes living in hilly areas where the food contains very little iodine cannot produce an adequate quantity of thyroid hormones. As a result, their metabolism rates are below normal, and this enhances the output of thyrotropin, which in turn stimulates thyroid gland in an attempt to produce increased quantities of thyroxine. Unfortunately, even this stimulus cannot increase the output of thyroxine when iodine is lacking. But the anterior pituitary gland continues to produce large amount of thyrotropin, so that the thyroid gland continues to enlarge, becoming progressively filled with colloid that contains almost no thyroxine. The enlarged gland is called endemic goitre. Thus, the patient with such goitre has functionally subnormal thyroid, but structurally or anatomically a large gland. In nodular nontoxic goiter the appearance of the thyroid gland vary greatly. There may be one or many asymmetrical nodular masses (multinodular goitre as shown in figure 3.9) formed in the gland. The development of hyperplasic nodules probably depends on the severity of the continuing iodine deficiency. As a precautionary step to control such disorder, table salt is compulsorily iodised in those areas where goitre is common.

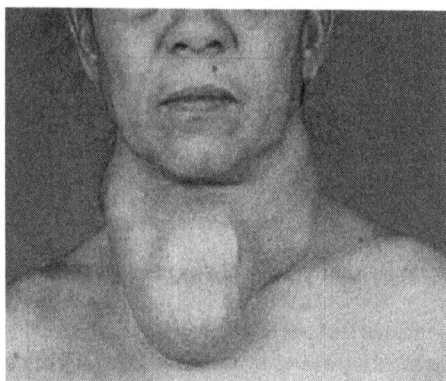

Figure 3.9 An asymmetrical multinodular goitre.

3.5.3 Thyroid Function Test

In conjunction with clinical evaluation of the patient, laboratory determinations are valuable in diagnosis and following the course of patients with hyper-and hypothyroidism. The patient's serum thyroid-hormone-binding capacity, age, drug therapy, and intercurrent illness can affect the results of various laboratory determinations of thyroid function.

Clinical judgment together with the patient's history and findings on physical examination is very important in correctly interpreting thyroid function test. In general, the most useful measurements to confirm or exclude hyperthyroidism are serum T_4, free T_4 (or free thyroxine index), and T_3; for hypothyroidism, they are serum T_4, free T_4 (or free thyroxine index), and TSH.

Estimation of circulating levels of thyroid hormones became possible when the protein bound iodine (PBI) method was developed. Small amounts of non-hormonal iodine are present in the serum of a healthy person under most clinical circumstances. 80 to 90 percent of PBI is derived from thyroxine. The PBI, which has a reference interval of about 4 to 8 μg/dl, is largely a reflection of serum T_4 concentration, because serum T_3 expressed in terms of iodine is only about 0.01 μg/dl. Serum T_4, by competitive protein binding or displacement analysis, is based on specific binding properties of thyroxine-binding globulin (TBG), thus allowing determination of T_4 independent of its iodine content. Serum T_4 by radioimmunoassay [T_4-(RIA)] method, employs an antibody to thyroxine. Advantages of serum T_4-(RIA) include its sensitivity and elimination of the extraction step. The reference interval for serum T_4 (RIA) in adults is approximately 5.5 to 12.5 μg/dl (72 to 163 nmol/l) expressed as thyroxine.

Serum T_3 levels usually are measured by immunoassay using highly specific T_3 antisera with little T_4 cross reactivity and a blocking agent such as sodium salicylate to eliminate endogenous T_3 protein-binding. The reference interval for serum T_3 is in the range of 60 to 160 μg/dl (0.92 to 2.46 nmol/l). Serum T_3 measurements may be helpful in confirming the diagnosis of hyperthyroidism, especially in patients with minimal elevation of serum T_4 or ambiguous clinical manifestation.

With changes in thyroxine-binding proteins, high or low levels of serum thyroxine occur, and these are not reflected in the corresponding alteration in the clinical state. In these situations, free (unbound) thyroxine is more closely related to the patient's clinical status. Free thyroxine may be quantitated using equilibrium dialysis or radioimmunoassay; the T_4 to TBG ratio (T_4/TBG) also has been employed to correct the changes in binding proteins. In addition, free thyroxine can be estimated using serum T_4 and T_3 uptake values to calculate the free thyroxine index (FTI).

Thyrotropin (thyroid stimulating hormone, TSH) usually is measured by radioimmunoassay; the TSH antibody used may cross-react with the glycoprotein hormones, giving falsely elevated values for TSH in such clinical situations as pregnancy (elevated human chorionic gonadotropin-hCG) and postmenopausal states (elevated FSH and LH). Sensitivity of TSH assays as well as their reference intervals vary from laboratory to laboratory; with sensitive TSH assay, mean value in healthy person is 1.5 mU/ml. TSH shows a circadian rhythm with maximum levels around midnight and minimum levels at noon.

TSH values following administration of thyrotropin releasing hormone (TRH) are used as a test of thyroid function. The TRH test may be helpful in individuals who have equivocal signs and symptoms of thyroid dysfunction but whose routine tests of thyroid function do not clarify the diagnosis. A TRH test may be performed in a number of ways which vary with the dose and route of TRH administration, as well as the parameter measured, but the intravenous test is most commonly used. Within about five minutes of intravenous administration, TRH causes a rise in serum TSH that reaches a peak in 20 to 30 minutes and returns to baseline in two to three hours. A typical response to intravenous injection of 500 µg of TRH is a rise in serum TSH to a concentration of about 16 to 26 µU/ml (16 to 26 mU/l) in women (slightly lower in male) from a mean value of 6 µU/ml (6 mU/l).

PARATHYROID GLANDS

These are in the form of four small, yellowish-brown coloured, oval bodies that are situated on the posterior surface of the thyroid lobes. Histologically, the majority of cells are *chief* and *water-clear* cells, which may represent differences in functional activity as a single cell type. With the age, increasing number of *oxyphil* cells appear; their cytoplasm is rich in eosinophilic granules shown by electron microscopy. During childhood, fat cells appear in the parathyroids and increase up to middle age.

Chief or principal cells are rich in glycogen and are mainly concerned with secretion of *parathyroid hormone* (PTH) or *parathormone*. It is a straight chain polypeptide hormone having 84 amino acids with a molecular weight of 9500. It is derived from a larger precursor, Pre-ProPTH, of 115 amino acids, which undergoes two successive cleavages both at the amino-terminal sequence to yield, first, an intermediate precursor, Pro-PTH, and then the hormone itself. Proteolytic enzymes inactivate PTH and it cannot be taken or given orally. The prime function of PTH is to regulate calcium metabolism.

3.6 BIOLOGICAL ACTION OF PARATHYROID HORMONES

1. PTH has vital importance, since it is essential for regulation of calcium and phosphate metabolism together with

vitamin D. This function can be achieved by acting on two organs that are bones and kidney. It enhances the absorption of calcium and phosphorus from the intestine thereby increasing the plasma calcium level and decreasing the phosphate level. If the parathyroid gland is removed from the body, serum calcium level decreases while the phosphorus level increases.

2. It is known to increase the serum calcium level by acting on the osteoblasts of bones and to promote excretion of phosphate by decreasing its reabsorption by the renal tubule and as a result serum phosphate level falls. Alkalosis or a raised plasma phosphate concentration may, by causing a fall in ionised calcium, also stimulate PTH secretion, and in chronic renal failure phosphate retention may result in hyperplasia of the parathyroid gland.

3. The action of PTH in kidney and bone is mediated through the stimulation of adenyl cyclase activity that ultimately leads to enhance cyclic AMP production. These effects precede and presumably mediate changes in phosphorous and calcium transport in kidney and bone.

4. PTH is essential for the normal process of eggshell formation in egg-laying animals. In the absence of PTH, soft-shelled eggs are produced for a while and finally the animal stops egg production.

3.7 CALCITONIN AND ITS BIOLOGICAL ACTION

It is a peptide hormone synthesized and secreted by specialized parafollicular or C-cells in the lateral lobes of the thyroid gland. There is evidence for a larger precursor form of calcitonin with a molecular weight of 15,000, which undergoes several cleavages to yield the major C-cell secretory product, calcitonin monomer of 32 amino acid residues with a molecular weight of 3500.

Calcitonin acts to lower the concentration of calcium in blood. Bone is continually remodeled through resorption of old bones and laying down new bone. Cells called osteoblasts break down bone and release calcium; osteoblasts, on the other hand, take up circulating calcium and deposit new bone. Calcitonin decreases the activity of osteoblasts and thereby shifts the balance of bone turnover to favour deposition of bone and removal of calcium from the blood. Because the turnover of bone in adult human is not very high, calcitonin does not play a major role in calcium homeostasis in adult humans. It is probably more important in young, growing individuals.

3.8 HORMONAL REGULATION OF CALCIUM AND PHOSPHORUS METABOLISM

3.8.1 Calcium Homeostasis

Calcium is the most abundant mineral element in the body that plays a vital role in the basic physiological processes like blood coagulation, neuromuscular conduction, maintenance of normal tone and excitability of skeletal and cardiac muscles, stimulation–secretion coupling in various exocrine glands, and preservation of cell membrane integrity and permeability, particularly in terms of sodium and potassium exchange. Maintenance of calcium homeostasis involves the participation of three major organs—the small intestine, the kidney, and the skeleton (figure 3.10). The mammary gland is also important

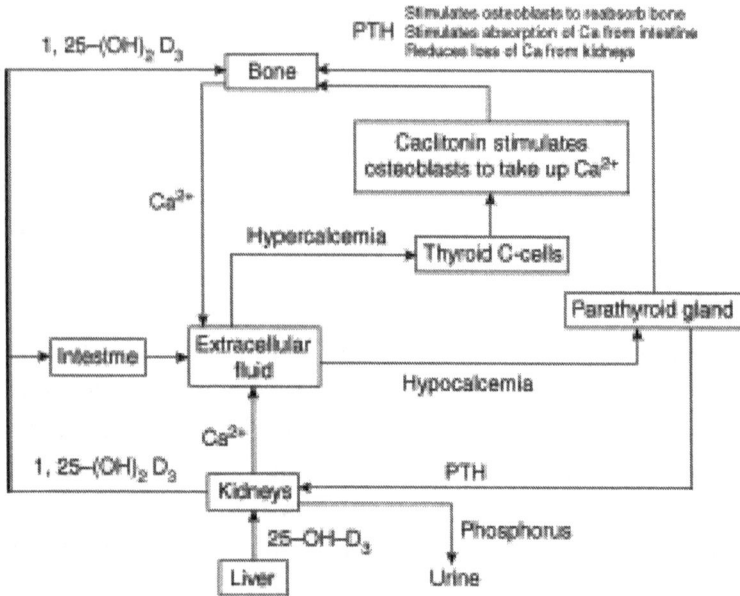

Figure 3.10   Hormonal regulation of calcium and phosphorus metabolism.

in maintaining calcium homeostasis during lactation, as are the placenta and foetus during pregnancy. Sweat glands are also responsible for small but significant excretion of calcium. As mentioned earlier there is no persistent net gain or loss of calcium in the adult human. During growth and pregnancy, a positive calcium balance must be maintained. The hormones that act principally upon major organs involved in calcium metabolism regulate calcium homeostasis. These hormones are parathyroid hormone and the hormones derived from renal metabolism of vitamin D_3, notably 1,25 dihydroxycholecalciferol [1,25-$(OH)_2D_3$]. Quite possibly calcitonin plays a role in the regulating process, although its significance in man is controversial. Other hormones like thyroid hormone, growth hormone, adrenal glucocorticoids and gonadal steroids have also an influencing role over the control of calcium metabolism.

It is commonly recommended that the daily dietary intake of calcium should be about 1200 mg during pregnancy and lactation and 800 to 1200 mg during childhood. Calcium is absorbed by active transport in the intestine. The major stimulant for calcium absorption is vitamin D. Absorption is also enhanced by growth hormone, an acid medium in the intestine, and increased dietary protein. The ratio of calcium to phosphorus in the intestinal contents is also important in that a ratio greater than two tends to inhibit calcium absorption because of the formation of insoluble calcium phosphate. Both cortisol and excessive alkalinity of the intestinal contents are inhibitory to calcium absorption.

The daily calcium excretion by the sweat glands widely ranges from 15 to 100 mg per day. The loss can be increased during extreme environmental conditions. The major loss of calcium is via urinary excretion, which accounts for 50 to 200 mg or more each day depending upon the dietary intake. Urinary calcium excretion is enhanced by hypercalcemia, phosphate deprivation, acidosis, and glucocorticoids while it is diminished by PTH, certain diuretics, and probably by vitamin D.

3.8.2 Phosphorus Homeostasis

Phosphorus plays a significant role as phospholipids, nucleic acids, nucleotides, constituent of cell membrane and cell cytoplasm, and compounds that are important in biochemical energy storage and exchange. Most of the phosphorus in extracellular fluid is inorganic and in the form of two species, HPO_4^{--} and $H_2PO_4^{-}$. The relative amount of the two phosphate ions are pH-dependent. At pH 7.4, the ratio of

HPO_4^{--} is about 4:1. In a healthy person, serum phosphorus varies in a range of 2.4 to 4.7 mg/dl. Higher phosphorus levels occur in growing children. Ingestion of food can significantly alter serum phosphorus concentration. Levels in adult females are lower than normal during menstruation.

Three major organs are concerned with phosphorus homeostasis: the small intestine, the kidney, and the skeleton, which functions as a storage reservoir. The average dietary intake for adults is about 800 to 1000 mg, most of which is derived from milk and dairy products. About two-thirds of ingested phosphate is absorbed in the small intestine. The unabsorbed phosphorus is excreted in the faeces. Intestinal absorption of phosphate is an energy-dependant active transport process. Absorption is increased in association with decreased dietary calcium and increased acidity of intestinal contents. Absorption is also accelerated by the action of vitamin D and growth hormone.

Most of the phosphorus is absorbed from the intestines of adults; the extra amount of phosphorus is excreted in urine (at the rate of about 0.35 to 1.0 gm of inorganic phosphorus per daily). About 90 percent of plasma phosphorus is filterable by the glomeruli. Ordinarily about 85 to 95 per cent of filtered phosphate is reabsorbed to maintain phosphorus balance. Parathyroid hormone inhibits renal tubular reabsorption of the phosphate.

3.9 PARATHYROID DYSFUNCTION

The secretory activity of the parathyroid gland can be altered in a variety of situations to cause hyperparathyroidism and hypoparathyroidism.

3.9.1 Hyperparathyroidism

The hyper functioning of parathyroid gland can be classified into primary and secondary hyperparathyroidism. Primary hyperparathyroidism is characterized by excessive secretion of PTH in the absence of an appropriate physiological stimulus and results in harmful effects. Secondary hyperparathyroidism occurs when the glands are exposed to increased stimulation, as in the case of kidney failure.

Primary hyperparathyroidism It is characteristically associated with elevated serum calcium (hypercalcemia) and decreased serum inorganic phosphorus and frequently accompanied by mild systemic acidosis. PTH acts directly on bones to cause increased resorption and consequent increase in serum calcium. Two other factors also contribute to elevated serum calcium; PTH stimulates increased renal biosynthesis of $1,25\text{-}(OH)_2\,D_3$, which increases intestinal absorption of calcium. PTH also promotes renal tubular reabsorption of calcium. The decreased concentration of inorganic phosphorus is primarily the result of PTH-induced phosphate diuresis caused by decreased renal tubular reabsorption.

The condition can occur at any age but is commonest in middle age and especially in post-menopausal women. The presenting features vary considerably depending on whether the symptoms and signs are mainly due to its effects on the kidney or bones or to hypercalcemia. The generalized symptoms are muscular weakness, tiredness, anorexia, thirst and polyurea. Clinically and pathologically significant feature is development of bone lesions producing the condition called *osteitis fibrosa cystica*. Increased osteoblastic activity leads to extensive destruction of bone tissue (*osteoporosis*), bone pain, skeleton deformities and fractures. Metastatic deposition of calcium can also occur in the walls of blood vessels, the lungs, gastric mucosa and renal tubules and the precipitation of calcium leads to formation of kidney stones (*nephrocalcinosis*).

Secondary hyperparathyroidism is characterized by an excessive secretion of PTH in response to chronic hypocalcemia. In most of the cases, chronic hypocalcemia is the result of either vitamin D deficiency or renal disease. Chronic renal failure can result in compensatory hyperparathyroidism, which in turn causes diffuse bone disease, including osteoporosis, osteosclerosis (areas of increased bone density), osteitis fibrosa cystica, and metastatic calcification. The disease complex is sometimes called renal osteodystrophy, or when it occurs in children, it is known as renal rickets.

3.9.2 Hypoparathyroidism

It is the result of parathyroidectomy or an unintentional consequence of thyroidectomy. Lack of parathyroid hormone leads to fall in serum calcium and a corresponding rise in inorganic phosphorus concentration. Hypocalcemia increases the excitability of sensory and motor nerves, with various effects depending on the severity. If severe, hypoparathyroidism results in tetany. The tone of skeletal muscles is increased and there may be spasm of the hands and feet in characteristic position, twitching and jerking movements and painful cramps of the limb muscles. There may also be generalized convulsions resembling epileptiform fits.

Idiopathic hypoparathyroidism is the rare acquired disorder that is common in women than men. It is an organ–specific autoimmune disease and is sometimes associated with other disorders like primary adrenocortical atrophy, pernicious anaemia, chronic thyroiditis or thyrotoxicosis.

Pseudohypoparathyroidism is a rare hereditary disorder characterized by signs and symptoms of hypoparathyroidism, skeletal defects and metastatic ossification. However, it is distinguished from true hypoparathyroidism in that plasma calcium concentration is low and plasma phosphorus is high in spite of an increased concentration of PTH in plasma. The disease is so called because the abnormalities are not responsive to parathormone, and the parathyroids are actually hyperplasic. The disorder is therefore not due to lack of PTH but failure to respond to it.

3.10 CALCITRIOL

Calcitriol and 1,25 dihydroxycholecalciferol are both scientific names for most active natural forms of vitamin D. The biochemistry of vitamin D is very different from that of other vitamins; vitamin D is actually a hormone.

3.10.1 Biosynthesis of Calcitriol

The story of vitamin D begins when a vitamin D precursor is eaten. The precursor we get from plants is called "ergosterol" and the precursor we get from eating animal tissues is called "7-dehydrocholesterol." These substances are absorbed into the body when they are eaten and transported to the skin for modification by sunlight radiation (hence the popular terminology of vitamin D as the "sunshine vitamin"). The animal origin substance is converted to what is called "vitamin D_3" or "cholecalciferol" while the plant substance becomes "Vitamin D_2" or " ergocalciferol." From here we will follow the animal origin hormone as it is the most metabolically active. The next stop is at the liver for modification (a hydroxyl group is added to the 25th carbon of the vitamin D_3 molecule) thus forming 25-hydroxycholecalciferol. After this, 25-hydroxycholecalciferol (often abbreviated as 25-D_3) circulates to the kidney for its final activation. Another hydroxyl group is added in the final activation to form 1, 25 dihydroxycholecalciferol (calcitriol).

3.10.2 Mechanism of Action of Calcitriol

Vitamin D regulates the blood levels of calcium and phosphorus, by increasing the absorption of these minerals and by promoting them into the bones. Vitamin D does this by stimulating DNA to produce transport proteins, which bind to the calcium and phosphorus and increase absorption through the small intestine. This stimulating feature is unusual for a vitamin and normally a function of hormones. Vitamin D also stimulates the uptake of these minerals by bone cells. This process is helpful in building strong bones and healthy teeth.

Calcitriol has several important actions. It enhances absorption of both calcium and phosphate from the gastrointestinal tract, promotes release of calcium and phosphate from the bones where they are stored, and causes the kidney not to excrete calcium and phosphate. In short, vitamin D is designed to increase the amount of calcium and phosphorus circulating in the bloodstream.

Parathyroid hormone is able to drive stored calcium and phosphorus from the bones, as is vitamin D so these hormones are able to work in concert here but in the kidney they have different functions. In the kidney, while vitamin D saves both calcium and phosphate, parathyroid hormone causes only calcium to be saved and phosphate to be dumped. Parathyroid hormone causes increased calcium and active vitamin D but reduced phosphorus in the bloodstream.

Calcitriol is able to elevate serum calcium in some cases, which can further destabilize the calcium–phosphorus balance and can create further *toxicity issues* to the kidney. Calcium is one of the parameters monitored in kidney failure patients. If calcium should elevate, calcitriol is discontinued for a week or so and calcium is rechecked. Due to the complex nature of the above hormone interactions, inadequate calcitriol can also cause calcium elevation.

Calcitriol cannot be used in patients who have an elevated (greater than 6 mg/dl) plasma phosphorus level. Calcitriol might elevate phosphorus further at this stage. In such patients, plasma phosphorus must be reduced by diet, fluid administration or by phosphate binders before calcitriol can be started.

It is helpful to monitor parathyroid hormone levels in patients on calcitriol therapy as some dosage adjustment is sometimes beneficial depending on the parathyroid hormone level. This kind of monitoring is not crucial to calcitriol use, however calcitriol can be given with or without food. Vitamin D deficiency causes *rickets* in children and *osteomalacia* in adults but less dramatic deficiencies contribute to osteoporosis. The condition leads to soft, fragile bones that are porous and break easily. Other deficiency symptoms include diarrhea, insomnia, nervousness, and muscle twitches.

3.10.3 Rickets

Vitamin D deficiency is particularly important in childhood and can be caused by inadequate dietary intake, intestinal malabsorption, or decreased synthesis of active metabolites as a consequence of inadequate exposure to sunlight. The usual child living in temperate climates receives too little sunlight during winter months to provide the amount of vitamin D needed for the absorption of calcium from the gut. Therefore, rickets usually develops in early spring. Depletion of calcium from bones in the absence of vitamin D leads to severe weakening of bones often results in fractures or deformed bones with decreased growth.

3.10.4 Osteomalacia

When vitamin D deficiency occurs in adults, it leads to osteomalacia. It is the disease caused by deficient mineralizaton of bone resulting from various disturbances in calcium and phosphorus metabolism due to absence of vitamin D. Including vitamin D deficiency, severe osteomalacia is the result of drastically decreased serum phosphate level than depletion of serum calcium concentration. As the disease progresses, the patient suffers from weakness, skeletal pain, deformities and fractures.

SYNOPSIS

❏   Thyroid gland is a bilobed structure situated in the neck.

❏   Histologically, human thyroid is composed of follicles filled with a colloidal material called *thyroglobulin*. It is an iodized glycoprotein containing *thyroxine* (T_4) and *triiodothyronine* (T_3).

❏   The parafollicular or 'C' cells present in interlobular space secrete the hormone called *calcitonin* or *thyrocalcitonin*.

O T_3 and T_4 raise basal metabolic rate and increase the O_2 consumption in most of the tissues. They are essential for normal physical and mental development and in the metabolism of protein, carbohydrate and fat.

O Secretion of thyroid hormone from thyroid gland is regulated by thyroid stimulating hormone (TSH) secreted by the anterior pituitary gland. The secretion of TSH is, in turn, regulated by thyrotropin-releasing factor (TRF) released from hypothalamus.

O The interrelationship of pituitary and thyroid is based on the principle of negative feedback mechanism.

O Excessive secretion of T_4 and T_3 causes a serious disorder known as hyperthyroidism or thyrotoxicosis that leads to Grave's disease or exophthalmic goitre while their inadequate secretion results in hypothyroidism that results in cretinism in children and myxoedema in adults.

O Nontoxic goitre (simple goitre) is the non-inflammatory condition of the thyroid gland that results in enlargement of the thyroid gland without hyperthyroidism.

O The most useful measurements to confirm or exclude hyperthyroidism are serum T_4, free T_4, and T_3 and for hypothyroidism, they are serum T_4, free T_4, and TSH. These measurements form the basis of the thyroid function test.

O Parathyroid glands are four small, yellowish-brown, oval bodies that are situated on the posterior surface of the thyroid lobes. Histologically, they are made of *chief* and *oxyphil* cells. *Chief* cells are concerned with secretion of *parathyroid hormone* (PTH) or *parathormone*.

O PTH enhances the absorption of calcium and phosphorus from the intestine and increases the plasma calcium level (by acting on osteoblasts of the bones) and decreasing the phosphate level (by promoting its excretion through kidney).

❍ Calcitonin is important in young, growing individuals where it acts to lower the concentration of calcium in blood by decreasing the activity of osteoblasts. Because the turnover of bone in adult human is not very high, calcitonin does not play a major role in calcium homeostasis in adult humans.

❍ Calcium and phosphorus metabolism are regulated by parathyroid hormone and the hormones derived from renal metabolism of vitamin D_3, notably 1,25 dihydroxycholecalciferol [1, 25-$(OH)_2$ D_3].

❍ Parathyroid function can be altered in a variety of situations to cause hyperparathyroidism and hypoparathyroidism. Its hyperfunctioning can be a type of primary and secondary hyperparathyroidism. Primary hyperparathyroidism is characterized by excessive secretion of PTH in the absence of an appropriate physiological stimulus. Secondary hyperparathyroidism occurs when the glands are exposed to increased stimulation, as in the case of kidney failure.

❍ Hypoparathyroidism is the result of parathyroidectomy or an unintentional consequence of thyroidectomy leading to tetany. Idiopathic hypoparathyroidism is the rare acquired disorder that is an organ-specific autoimmune disease while pseudohypoparathyroidism is a rare hereditary disorder characterized by signs and symptoms of hypoparathyroidism.

❍ Calcitriol is the most active natural form of vitamin D also known as 1, 25 dihydroxycholecalciferol [1, 25-$(OH)_2$ D_3]. It increases the amount of calcium and phosphorus circulating in the bloodstream.

❍ Vitamin D deficiency causes *rickets* in children and *osteomalacia* in adults that are characterized by weakness, skeletal pain, deformities and fractures.

REVIEW QUESTIONS

1. Explain the structural features of thyroid gland.

2. Describe the biosynthesis of thyroid hormones.

3. Enlist the biological functions of T_4 and T_3 hormones.

4. What is negative feedback mechanism? Explain it in relation to thyroid and pituitary gland.

5. What is thyrotoxicosis? How does exophthalmic goitre differ from simple goitre?

6. Explain the term hypothyroidism. Describe the symptoms of cretinism and myxoedema.

7. Describe thyroid function test with its significance.

8. Explain the structural features of parathyroid glands.

9. Enumerate the role of various hormones in calcium and phosphorus metabolism.

10. List the points of differences between primary and secondary hyperparathyroidism.

11. Describe the disorders caused by hypoparathyroidism.

12. Give the importance of calcitriol with its biosynthesis and mechanism of action.

13. Write short notes on:
    i. Tyrosine
    ii. Parafollicular cells
    iii. Rickets
    iv. Calcitonin
    v. Vitamin D

vi.   Exophthalmos

vii. PTH

viii. Ergocalciferol

ix.   TRF

x.   Tetany

FOUR

ADRENAL GLAND

4.1 STRUCTURAL FEATURES

Adrenal or suprarenal gland is a pair of glands located immediately above each of the two kidneys in the form of flattened structures (figure 4.1). The average weight of the gland is about 5–9 g in the adult. Structurally, functionally and embryologically, each adrenal gland consists of a gland within a gland. The outer part is called adrenal cortex that is derived from mesoderm while inner part is known as adrenal medulla, which is derived from ectoderm or neural crest. These glands produce hormones of two entirely different types. Adrenal cortex secretes mineralocorticoids, glucocorticoids, and androgens, while adrenaline and noradrenaline are the hormones secreted by adrenal medulla and their secretion is directly under the influence of sympathetic nerves.

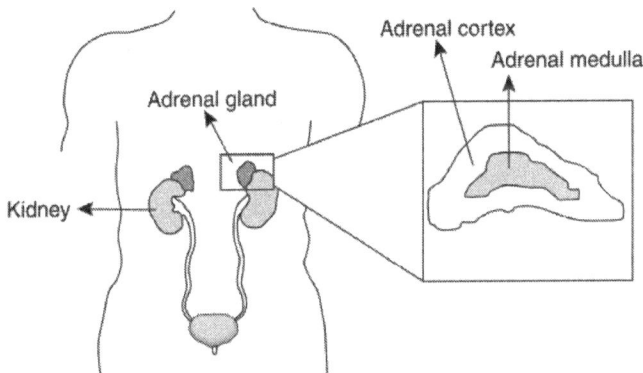

Figure 4.1 Adrenal gland showing outer cortex and inner medulla.

4.2 HORMONES OF ADRENAL CORTEX

The outer portion of the adrenal gland is the cortex that is composed of large, fat-laden epithelial cells arranged in three zones.

1. *Zona glomerulosa* in the form of outer thin layer of irregularly arranged cells that secrete *mineralocorticoids.*

2. *Zona fasciculata* is the widest zone having radially arranged strands of cells in form of fascicles, which secrete *glucocorticoids.*

3. *Zona reticularis* is the layer of the adrenal cortex that is composed of irregularly arranged cells leaving wide blood spaces. The cells of this zone synthesize sex hormones, *androgens.*

The various zones of adrenal cortex secrete different adrenocortical hormones, therefore the zones are structurally and functionally different. However, hormones secreted by adrenal cortex are all chemically similar since they are all steroids derived from cholesterol and are collectively called *corticosteroids.*

Synthesis of adrenocortical hormones   The simplified pathways of adrenocortical hormones are shown in figure 4.2,

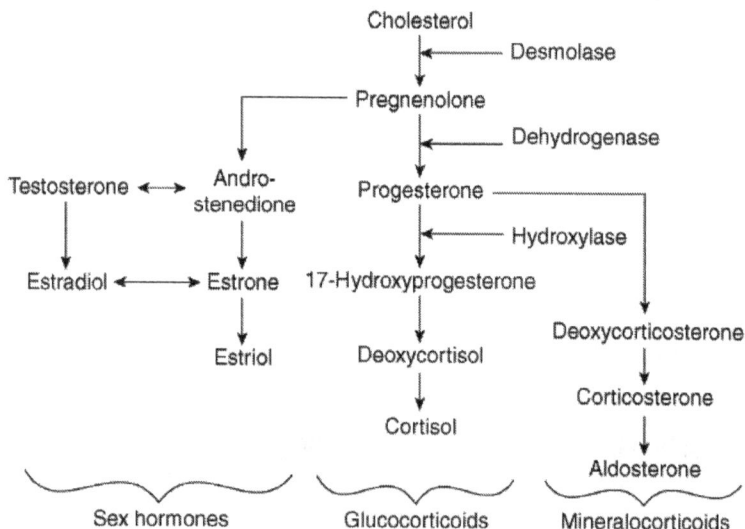

Figure 4.2 Simplified pathways for synthesis of adrenocortical hormones.

in which four general types of enzymes are involved—hydroxylases, dehydrogenases, desmolases and isomerases. These enzymes are named depending on the carbon atom transformed. Cholesterol acts as a precursor for all steroid hormones. Desmolase breaks down the side chain of cholesterol converting it to pregnenolone, which in turn is acted upon by various enzymes to synthesize different corticosteroids.

4.2.1 Mineralocorticoids

These are secreted by the zona glomerulosa of adrenal cortex in three different forms—*aldosterone, corticosterone,* and *deoxycorticosterone.* These hormones are mainly concerned with regulation of mineral metabolism mainly sodium and potassium. If the adrenal glands are removed from the body of an animal, it must have sodium added to its diet, or its sodium will be depleted and it will die. One human patient with a nonfunctional adrenal gland compensated by salting her food heavily and, in addition, ate 60-pound block of salt in the course of a year. Out of three mineralocorticoids, aldosterone is the most active and performs the following functions

1. It promotes sodium reabsorption in the renal tubules causing increase in the sodium chloride content of serum, lymph, and tissue fluid. To achieve this, various enzymes are required and since enzymes are the products of gene expression, it is believed that aldosterone activates the genetic mechanism of the tubular epithelial cells to increase the production of specific enzymes required for sodium absorption.

2. It decreases the rate of reabsorption of potassium in the renal tubule thereby causing increased excretion

of potassium through kidneys and fall in the serum potassium level.

3. Since aldosterone causes retention of sodium in the blood and tissues, this leads to rise in osmotic pressure, retention of water in the body and increase in arterial pressure. On the other hand, a decrease in the blood volume from dehydration or sodium deficiency brings about the drop in blood pressure that stimulates the juxta-glomerular cells of the kidney and through the *renin-angiotensin mechanism*, the blood pressure is brought to normal. The details of the mechanism are as shown in the figure 4.3.

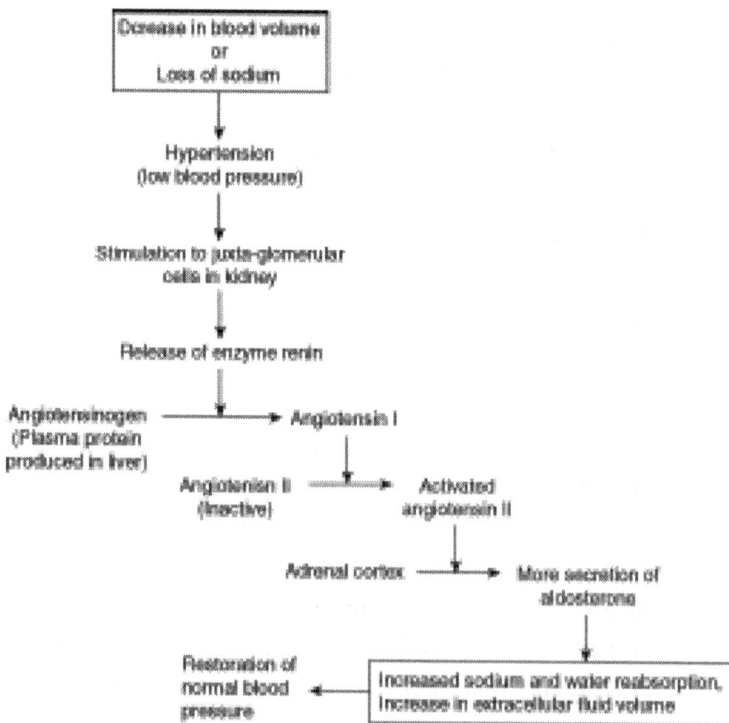

Figure 4.3 Role of aldosterone in restoration of blood pressure.

Regulation of aldosterone secretion The secretion of aldosterone from zona glomerulosa is regulated by the levels of sodium and potassium in the blood of an animal. Increase in concentration of sodium in serum inhibits the secretion of aldosterone. On the other hand, a lack of sodium in the blood and subsequent changes in the blood volume and blood pressure influences the secretion of the hormone. In addition, the most potent long-term stimulator of aldosterone secretion in experimental animals is a rise in potassium ion concentration in the plasma. The increased secretion of aldosterone in turn causes the kidneys to excrete large amounts of potassium from the body, thereby reducing the potassium concentration in the extracellular fluid to normal.

Adrenocorticotropin or ACTH secreted by anterior pituitary gland also causes a slight to moderate increase in the secretion of aldosterone but this is not enough to have major significance.

4.2.2 Glucocorticoids

These are the hormones secreted by zona fasciculata of the adrenal cortex under the influence of ACTH from adenohypophysis. They are of three types, namely hydrocortisone, corticosterone, and cortisone. Of these hormones, hydrocortisone or cortisol is the most active form that shows at least 95% of the glucocorticoid activity. Their names suggest that they have significant role to play in metabolism of carbohydrates, fat and protein. The major physiological functions carried out by glucocorticoids are as follows:

1. *Effect on carbohydrate metabolism*
   a. They promote glycogen formation in the liver and muscles (glycogenesis),

b.   Rise blood sugar level (hyperglycemia),

c.   Increase gluconeogenesis in the liver (proteins and glycerol portion of fats are converted into glucose),

d.   Decrease glucose uptake by tissues,

e.   Increase absorption of glucose from the intestine.

2.   *Effect on protein metabolism*

a.   Accelerate the rate of deamination and breakdown of tissue proteins to amino acids,

b.   Promote conversion of proteins into glucose (gluconeogenesis) in liver

c.   Decrease the quantity of protein in most tissues of the body.

3.   *Effect on fat metabolism*

a.   Decrease in the amount of fat in the storage tissues and increased use of fat for production of energy,

b.   Stimulate absorption of lipids from the intestine.

c.   If immobilized fat is not used immediately for energy formation, their concentration in extracellular fluid can increase to cause acidosis.

4.   Glucocorticoids are known as anti-inflammatory hormones since they have the ability to suppress the process of inflammation.

5.   *Effect on lysosome activity* When the cells are damaged due to trauma or because of disease, the lysosomes break open to release their hydrolytic enzymes in the cytoplasm, which in turn autolyze internal structures of the cells, decrease the cellular function or actually kill the cells. Therefore, it is very important to prevent this process. Cortisol has the special ability to stabilize

the membranes of lysosomes thereby preventing their break off. Fortunately, in a person with certain disease or trauma, large quantity of cortisol are secreted in the blood that protect a person from dangerous effects of some diseases by preventing the rupture of the lysosomes inside the cells.

6. Administration of corticosteroid in an animal causes decrease in number of eosinophils (eosinopenia) and lymphocytes (lymphopenia).

7. Glucocorticoids have an influencing role in the excitability of the brain and on the metabolism of the nervous tissue.

8. Cortisol critically mediates the body's response to stress. It protects the body from stressful situations that may be the result of fright, strokes, ulcer, cardiovascular problems, impaired immune system function and increased susceptibility to cancer and other diseases.

Regulation of cortisol secretion   Cortisol secretion is controlled by ACTH from anterior pituitary gland, which in turn is controlled by corticotropin–releasing factor (CRF) from the hypothalamus. Because the cortisol response to a stressor includes, secretion, diffusion, circulation, and cell activation, it is much slower than the epinephrine response. Also, many of the actions of cortisol involve changes in the gene expression and that takes time. The primary stimulus for glucocorticoid secretion is the 'stress' that results from the fright, strokes, or from any other painful condition in the body. The stress probably causes the chain of reactions by initiating the nerve impulses that are transmitted from the periphery into the hypothalamus. In response to this, hypothalamus secretes the substance CRF, which passes through the hypothalamic-hypophyseal portal system into adenohypophysis. In turn, anterior pituitary gland secretes corticotropin that flows in the blood to adrenal cortex where it elicits cortisol secretion. Then, the cortisol promotes gluconeogenesis, mobilizes protein and fat, and stabilizes the lysosomes. Thus, prepares a person to face the stressful situation and also help in the repair of the damage (figure 4.4).

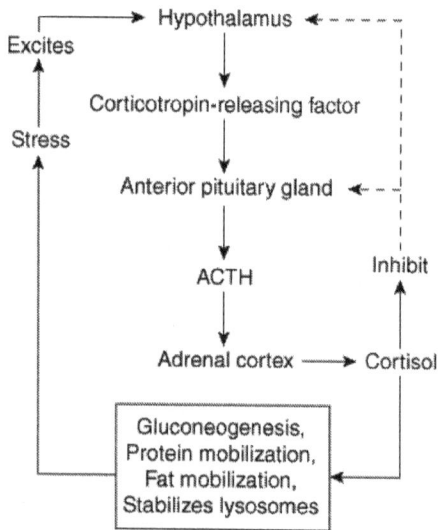

Figure 4.4   Mechanism for regulation of cortisol secretion.

Turning off the cortisol response is as important as turning it on. A study of stress in rats showed that old rats could turn on their stress responses as effectively as young rats, but that had lost the ability to turn them off as rapidly. As a result, they suffered from the well-known consequences of stress seen in humans. Further research showed that the stress responses are turned off by negative feedback action of cortisol on cells in the brain that causes a decrease in release of CRF. Repeated activation of this negative feedback mechanism, either through repeated stress or through prolonged medical use of cortisol, leads to gradual loss of cortisol-sensitive cells in brain, and therefore to a decreased ability to terminate stress response.

4.2.3 Adrenal Cortex Sex Hormones

These are ketosteroids mainly secreted by the zona reticularis of the adrenal cortex. Their physiological role is obscure in adults. The quantities of androgen and oestrogen secreted by adrenals are insignificant compared with the endocrine activity of the gonads. In pathological conditions, excessive secretion of adrenal sex hormones can cause premature puberty, virilism in females and very rarely feminisation in male.

4.3 ABNORMALITIES OF ADRENOCORTICAL SECRETION

As stated earlier, adrenal cortex secretes three types of steroid hormones, glucocorticoids, mineralocorticoids and sex hormones. And therefore there are three corresponding types of adrenal cortical hyperfunction.

4.3.1 Cushing's syndrome (Chronic Hypercotisolism)

The disorder results from excessive secretion of cortisol by the adrenal cortex or a cortical tumour. Although its natural occurrence is uncommon, it is important to recognize it as a potentially fatal condition that requires immediate treatment. Similar features are observed in patients who are on prolonged treatment with high doses of glucocorticoids or ACTH.

Figure 4.5   A patient with Cushing's syndrome.

Cushing's syndrome occurs most often in women over a wide age range but is also seen in men and children. In all cases the main features are due to excessive cortisol secretion although there is frequently in addition excessive secretion of androgen that results in virilism. The clinical features include:

1. painful adiposity of the face, neck and trunk which contrasts with the relatively thin limbs (figure 4.5);

2. increased protein breakdown, with consequent thinning of the skin, generalized osteoporosis, and wasting and

weakness of the skeletal muscles. Fat accumulates in the abdomen, resulting in purple striae in which small vessels are visible through the stretched skin;

3. systemic hypertension;

4. diminished glucose tolerance, sometimes a very high blood sugar level leads to adrenal diabetes;

5. an increased tendency to bacterial infections and poor healing of wounds, and

6. arrest of growth in affected children.

Unless treated, Cushing's syndrome is likely to result in death from the effects of hypertension, from bacterial infection or from suicide.

Secondly, Cushing's syndrome may be due to excessive secretion of ACTH by the adenohypophysis. This form of the Cushing's syndrome is sometimes confusingly called Cushing's disease.

4.3.2 Addison's Disease (Chronic Adrenocortical Insufficiency)

Primary adrenocortical insufficiency (Addison's disease) usually occurs from an autoimmune process or from tuberculosis and other granulomatous diseases. When hypocortisolism results from a pituitary lesion, it is called secondary adrenal insufficiency. Patients with primary adrenal insufficiency have deficiencies of both glucocorticoids and mineralocorticoids in contrast to individuals with secondary adrenal insufficiency, who have only a glucocorticoids deficiency. Addison's disease is a disorder caused by hyposecretion of adrenocortical hormone. Its outstanding features are weakness, loss of appetite and weight, and hypotension. Libido is usually diminished and the skin becomes pigmented, particularly the exposed parts and the external genitalia.

In the absence of high salt intake or steroid therapy, failure of reabsorption of salt in the renal tubules results in *hyponatraemia* that is accompanied by chronic dehydration. These features result from the combined deficiency of mineralocorticoids and glucocorticoids. Mineralocorticoid deficiency results in loss of salt and water, with consequent hypotension, weakness and some weight loss. Chronic glucocorticoid deficiency contributes to mental confusion, hypoglycemia, increased secretion of ACTH and MSH and poor reaction to stress. Anorexia, nausea, vomiting, weight loss and increased chloride deficiency, are also due to lack of glucocorticoids.

4.3.3 Congenital Enzyme Disorders of the Adrenal

There are at least eight different metabolic defects in synthesis of cortisol and aldosterone, each have been recognized by a deficiency of a specific adrenal enzyme. These enzymatic deficiencies all are inherited as autosomal recessive traits with variable degree of penetrance. Depending on the severity and location of the enzymatic defect in the metabolic pathway, deficiency of glucocorticoids, mineralocorticoids and sex hormones occurs, resulting in shock, salt wasting, or anomalous sexual development. Other findings such as hypertension, due to accumulation of mineralocorticoids, or virilization, which occurs from shunting of metabolism toward the sex hormone pathway, are clinically useful in differentiating the various enzyme deficiencies. Due to decreased cortisol production, the usual feedback mechanism is interrupted in these syndromes and there is a compensatory increase in ACTH, the magnitude of which may be sufficient to prevent adrenal insufficiency. This compensatory increase in ACTH leads to hypersecretion of steroid precursors and is responsible for development of hyperplastic adrenal gland.

4.4 HORMONES OF ADRENAL MEDULLA

Adrenal medulla is composed of irregularly arranged strands of chromaffin cells that are separated by sinusoidal vessels.

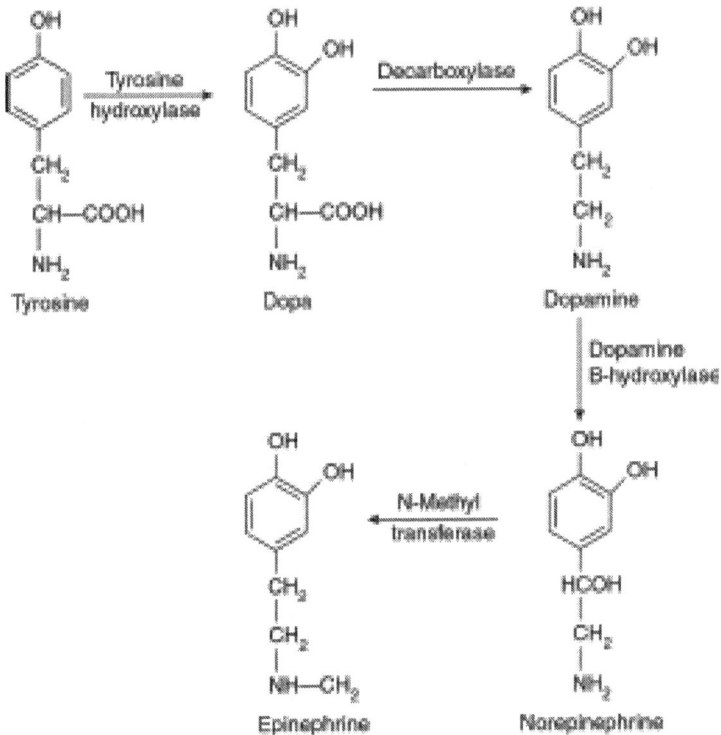

Figure 4.6 Biosynthetic pathways for catecholamines.

Medulla is richly supplied with a network of sympathetic nerves. The hormones of adrenal medulla are epinephrine (adrenaline) and norepinephrine (noradrenaline); both are called catecholamines. The major catecholamine biosynthetic pathways are shown in figure 4.6 and that involve amino acid tyrosine as precursor for synthesis of catecholamines. Tyrosine undergoes hydroxylation to form Dopa, (Dihydroxyphencyalanine), which is decarboxylated to Dopamine and subsequently to norepinephrine and finally to epinephrine. Conversion of norepinephrine to epinephrine occurs mainly in the adrenal medulla. In the central nervous system, neurons produce dopamine and norepinephrine that act as neurotransmitters.

4.4.1 Biological Action of Adrenaline and Noradrenaline

The secretory cells of adrenal medulla are actually the modified postganglionic sympathetic neurons, and these cells,

Table 4.1 Effect of catecholamines on various systems.

System	Function	Epinephrine	Norepinephrine
1. Cardiovascular System	Heart rate	Increases	Slightly increases
	Cardiac output	Increases	No change
	Systolic B.P.	Increases	Increases
	Diastolic B.P.	No change	Increases
	Blood vessels	Dilates in muscles & constricts in others	Almost constricts
2. Respiratory System	Bronchi	Dilates	Dilates

3. Skin	Blood flow	Decreases	Decreases
	Hair	Pilo erection	Pilo erection
	Sweat gland	Induces sweating	Induces sweating
4. Eye	Pupil	Dilates	Dilates
5. Excretory System	Renal tubules	Constricts	Constricts
	Urine volume	Reduces	Reduces
6. Central nervous System	Brain	Increases anxiety	No effect
	Adenohypophysis	Stimulates	Stimulates
7. Digestive System	Salivary gland	Stimulates	Stimulates
	Liver	Mobilizes glycogen	Mobilizes glycogen
	Intestine	Stimulates	Inhibits
8. Metabolism	B.M.R.	Increases	Increases
	Oxygen consumption	Increases	Increases
	Blood sugar	Increases	Increases

when stimulated, secrete epinephrine and norepinephrine into the blood. These two hormones then circulate in the blood and are distributed to the cells of the body. The circulating norepinephrine has the same stimulatory or inhibitory effects as the norepinephrine released directly by the sympathetic nerves. And except a few instances, the epinephrine has almost the same effects (table 4.1). Therefore, these two hormones in combination produce almost the same effects on all the organs of the body as direct sympathetic nerve stimulation.

4.4.2 Emergency Hormones

The adrenal medulla produces catecholamines in response to emergency conditions that disturb the physical and mental state of the body. The stressful situations due to *flight, fright, or fight* stimulate the adrenal medulla to secrete epinephrine and norepinephrine so that the body can handle the situation and subject can arouse for the action by (a) increasing heart rate and blood pressure, (b) diverting the blood flow to active muscles, (c) mobilizing liver glycogen, (d) increasing blood glucose level, (e) dilating bronchioles to help pulmonary ventilation and (f) increasing basal metabolic rate. The physically threatening events, such as encountering a mugger, or by events that are mentally stressful, such as giving a public speech or taking a test, or event the physical exercise, sudden exposure to cold, and the conditions like hypotension, asphyxia, cerebral anaemia, hypoglycemia, anesthesia, etc. can cause increased secretion of catecholamines from adrenal medulla.

Action of catecholamines on receptors Epinephrine and norepinephrine are both water-soluble amine hormones that bind to the same receptors on the surface of target cells. These receptors can be grouped into two general types, α-adrenergic and β-adrenergic receptors, which stimulate different actions within cells. Epinephrine acts equally on both types, but norepinephrine acts mostly on α-adrenergic receptors. The binding of catecholamines on α and β receptors has profound influence on the function of many type of cells, and also provides opportunity to develop drugs that activate or block α and β receptors. Stimulation of β receptors activates adenyl cyclase activity and so increases the intracellular cyclic AMP, which in turn initiates cellular responses. There are drugs called *beta-blockers,* which selectively block β-adrenergic receptors, can reduce the fight-or-flight responses to epinephrine without disrupting the physiological regulatory functions of norepinephrine. Beta-blockers like Alprenolol, Propranolol, Acebutolol, Atenolol, Oxprenolol, Bisoprolol, etc.

are commonly prescribed by physicians to reduce stress. Actually, these drugs block the β-effect of adrenaline on receptors on heart muscle cells and help to minimize cardiovascular disorders such as angina pectoris, hypertension, cardiac failure, myocardial infarction, obstructions of ventricular outflow, and also reduce the problems related to central nervous system like anxiety and migraine.

Regulation of catecholamine secretion   The secretion of catecholamines from adrenal medulla is exclusively controlled by the sympathetic nervous system. It is experimentally proved that stimulation of the splanchnic nerve, which innervates adrenals, initiates secretion of catecholamines while cutting this nerve leads to decreased secretion. The nerve centre, controlling the secretory activity of chromaffin cells of adrenal medulla and sympathetic ganglia is situated in the

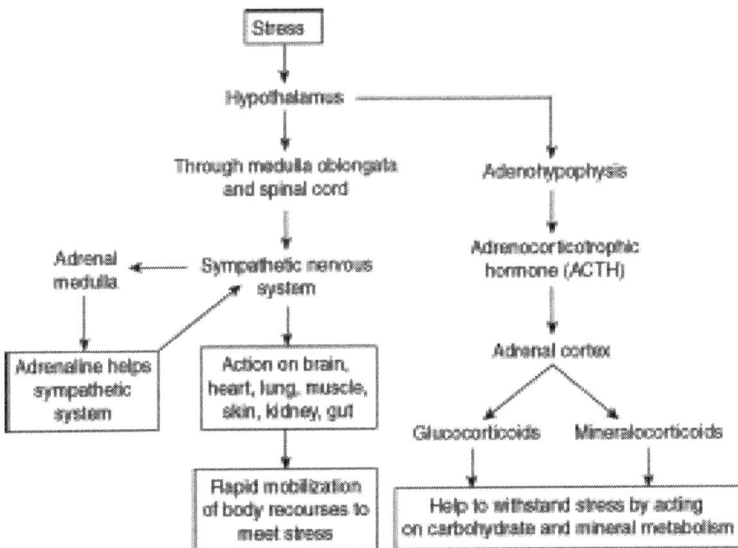

Figure 4.7 Cascade of events occurring in the body to meet a stressful situation.

hypothalamus. Stressful stimuli like pain, injury, cold, heat, etc., excite this centre, which in turn stimulates the sympathetic nervous system. Excitatory impulses from sympathetic nerves reach chromaffin cells where they bring about the secretion of adrenaline and noradrenaline (figure 4.7).

4.5 DISORDER DUE TO TUMOUR OF ADRENAL MEDULLA—PHAEOCHROMOCYTOMA

Hypersecretion of catecholamines is observed in patients with a functioning tumour of the adrenal medulla or of chromaffin cells elsewhere, the diseased condition called *phaeochromocytoma*. Its clinical symptoms are excessive sweating, nervousness, tremors, psychoses, attacks of blanching or flushing, tachycardia, headache, palpitations, loss of appetite and reduction in weight. In addition, the fasting blood sugar is usually raised and the basal metabolic rate increased. Obviously a phaeochromocytoma can mimic various other disorders. In most cases the diagnosis can be confirmed by demonstrating that the daily urinary excretion of vanillin mandelic acid (VMA, a metabolite of catecholamine) is more than doubled. The response to the α-blocking agent phentolamine is also of diagnostic value.

Although phaeochromocytoma is rare, its diagnosis and successful treatment are of great importance, because it causes a potentially curable form of hypertension, which may be fatal if left untreated. Early surgical removal of the tumour relieves the symptoms.

SYNOPSIS

O Adrenal or suprarenal gland is situated above the kidney and is divided into an outer part adrenal cortex and inner part adrenal medulla. Both parts produce hormones of two entirely different types. The cortex secretes mineralocorticoids, glucocorticoids, and androgens. Adrenaline and noradrenaline are the hormones secreted by adrenal medulla and their secretion is directly under the influence of sympathetic nerves.

O Hormones secreted by adrenal cortex are steroids, derived from cholesterol and are collectively called *corticosteroids*.

O Hormones of adrenal medulla are derivatives of the amino acid tyrosine and are called *catecholamines*.

O Mineralocorticoids are secreted by zona glomerulosa of adrenal cortex in three different forms—*aldosterone, corticosterone, and deoxycorticosterone* that are mainly concerned with regulation of mineral metabolism.

O Aldosterone-mediated renin-angiotensin mechanism helps to maintain the blood pressure.

O Glucocorticoids are the hormones secreted by zona fasciculata of the adrenal cortex in the form of hydrocortisone, corticosterone, and cortisone.

O Glucocorticoids play a significant role in the metabolism of carbohydrates, fat and protein. In addition, they are anti-inflammatory hormones and mediate the body's response to the stress besides the lysosomal stabilizing effect.

O Cortisol secretion is controlled by ACTH from adenohypophysis, which in turn is controlled by corticotropin-releasing factor (CRF) from the hypothalamus.

O The quantities of sex hormones secreted by adrenal cortex are insignificant.

O *Cushing's syndrome* (chronic hypercotisolism) is the disorder that results from excessive secretion of cortisol by the adrenal cortex or a cortical tumour. Its clinical features include painful adiposity of the face, neck and trunk with relatively thin limbs, osteoporosis, wasting and weakness of the skeleton muscles, accumulated fat in the abdomen showing purple striae, systemic hypertension, diminished glucose tolerance, high blood sugar level leading to adrenal diabetes, poor healing of wounds, and arrested growth in affected children.

❍ *Addison's disease* (chronic adrenocortical insufficiency) is due to an autoimmune process or from tuberculosis and other granulomatous diseases. Its features are weakness, loss of appetite and weight, hypotension, reduced libido and pigmented skin.

❍ The enzymatic defect in the metabolic pathway, deficiency of glucocorticoids, mineralocorticoids and sex hormones leads to shock, salt wasting, or anomalous sexual development.

❍ The stressful situations due to flight, fright, or fight stimulate the adrenal medulla to secrete catecholamines to handle the situation and arouse the body for the action by increasing heart rate and blood pressure, diverting the blood flow to active muscles, mobilizing liver glycogen, increasing blood glucose level, accelerating ventilation and increasing basal metabolic rate.

❍ The binding of catecholamines on α and β receptors on the cell has profound influence on the function of many type of cells, and also provided opportunity to develop drugs that activate or block α and β receptors. The drugs called *beta-blockers,* which selectively block β-adrenergic receptors, can reduce the flight-or-flight responses.

❍ The secretion of catecholamines from adrenal medulla is exclusively controlled by sympathetic nervous system.

❍ *Phaeochromocytoma* is due to hypersecretion of catecholamines in patients with a functioning tumour of the adrenal medulla or of chromaffin cells elsewhere. Its clinical symptoms are excessive sweating, nervousness, tremors, psychoses, attacks of blanching or flushing, tachycardia, headache, palpitations, loss of appetite and reduction in weight. In addition, the fasting blood sugar is usually raised and the basal metabolic rate increased.

REVIEW QUESTIONS

1. Describe the structure of adrenal gland.

2. Explain how adrenal medulla is different from the adrenal cortex in the structure and function.

3. What are the different mineralocorticoids? Explain their role in mineral metabolism.

4. Explain the role of glucocorticoids in regulation of various metabolisms.

5. Discuss the important factors that regulate aldosterone secretion.

6. What is the relationship between aldosterone and renin-angiotensin system?

7. Review the effects of hypo- and hypersecretion of corticosteroids.

8. Explain the roles of adrenaline and noradrenaline.

9. How does adrenal medulla help in stressful situation?

10. Write short notes on:

    i.   Adrenal sex hormones

    ii.   Norepinephrine

    iii. Role of CRF in release of corticosteroids

    iv.   Lysosome stabilization

    v.   Cortisol in stress

    vi.   Beta-blockers

    vii. Phaeochromocytoma

    viii. Adrenal medulla and sympathetic nervous system

    ix.   Emergency hormones

FIVE

ENDOCRINE PANCREAS

The pancreas is a dual-function organ situated in the loop of stomach and duodenum. Its exocrine part synthesizes and releases digestive enzymes in the intestine while the endocrine portion is made up of islets of Langerhans that secrete two important hormones, insulin and glucagon directly into the bloodstream.

The prominent function of the pancreas is that of an exocrine gland, that is histologically composed of pancreatic *acini,* which are serous in character and are lined by cuboidal epithelium. Acini synthesize and drain their secretion in the form of pancreatic juice into the duodenum through the pancreatic duct and help for the digestion of protein, fat and carbohydrate present in food.

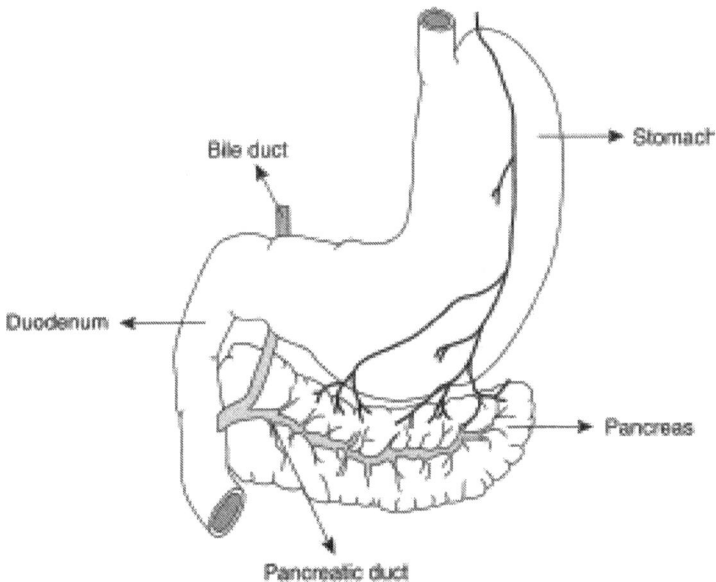

(a)

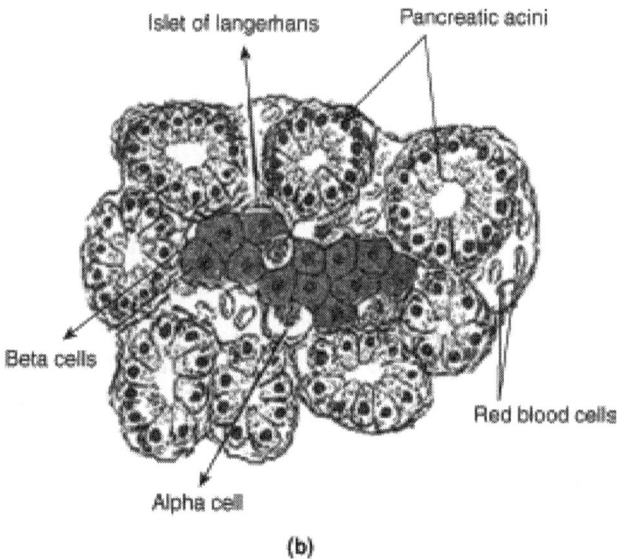

Figure 5.1 (a) Pancreas with its associated organs, (b) histological structure.

5.1 ISLETS OF LANGERHANS

The endocrine pancreas consists of the little collection of different types of the cells that are distributed through the substance of pancreas. These cells form islets of Langerhans, which contains three types of the cells, the alpha (α), the beta (β), and delta (δ) cells. Their special staining properties and presence of granules make them distinguishable from one another. In human pancreas, there may be as many as 2 million islets. About 65% of the total cells found in islets of Langerhans are beta types that synthesize the hormone, *insulin*. The alpha cells are about 25% of the total cells, which secrete the hormone *glucagon*. And 5 to 10% of these cells are the delta cells that secrete the hormone *somatostatin*. In addition to these cells, there is a one other type of cells known as PP cells that secrete a hormone called *pancreatic polypeptide* whose function is obscure. The hormones secreted by islets of Langerhans are directly poured into the circulating blood.

5.2 INSULIN

It is a small protein hormone secreted by the beta cells of islets of Langerhans. In 1922, Banting and Best for the first time isolated insulin from the pancreas. It is composed of two polypeptides, A and B chains that are held together by disulphide bonds. The chain A is composed of 30 amino acid residues while chain B contains 21 amino acid residues, thus in all insulin has 51 amino acid residues with a molecular weight of 6000 daltons.

5.2.1 Biosynthesis of Insulin

The synthesis of hormones is either a result of direct transcription and translation of the gene or the product of the gene in the form of an enzyme that mediate the synthesis of a hormone. Insulin is a small peptide that is formed directly by the genetic translation via messenger RNA. A polypeptide precursor of insulin, called *proinsulin,* is synthesized in the microsomal fraction of the pancreatic beta cells as a long single chain with a molecular weight of 9000 daltons. As a result of post-translational change, the proinsulin is converted to a double-chain molecule by the proteolytic enzyme treatment that removes the 31 amino acid C-peptide, thus forming insulin (figures 5.2 and 5.3).

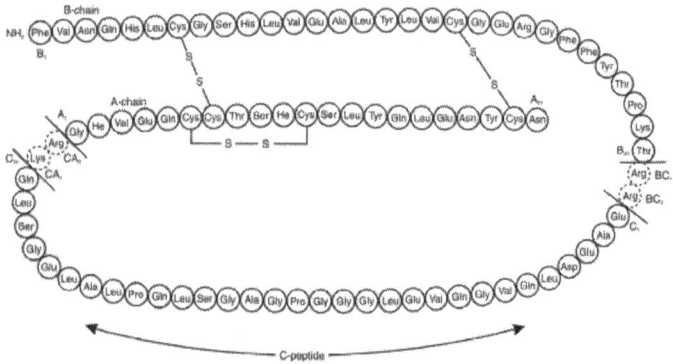

Figure 5.2 Proposed amino acid sequence of proinsulin. "C" peptide along with BC_1-BC_2 and CA_1-CA_2 will be removed to form insulin with A and B chains.

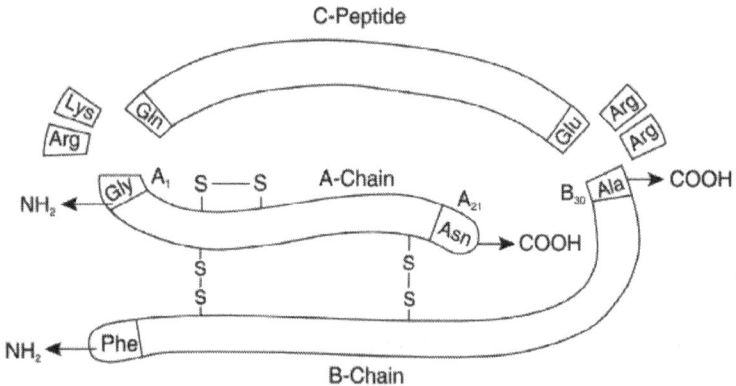

Figure 5.3 Proinsulin is converted to insulin by enzymatic cleavage.

Similar to many protein hormones, insulin is stored in the granules that originate from the Golgi apparatus present in the pancreatic beta cells. The precursor, or proinsulin, becomes associated with the granules and it seems likely that its conversion to insulin takes place in these. The granules can travel to the peripheral regions of the cells, and, in response to the releasing-stimuli i.e. when blood glucose level rises, glucose enters the pancreatic β cells where it undergoes oxidative phosphorylation to form ATP that helps to close the ATP-gated K^+ channel in the plasma membrane. This leads to depolarization of the membrane, which in turn, helps to open the voltage-gated Ca^{++} channel to open and allows the Ca^{++} ions to enter the cell. As a result, the cytoplasmic concentration of Ca^{++} ions increases and this is enough to release the insulin from the cell by exocytosis (figure 5.4).

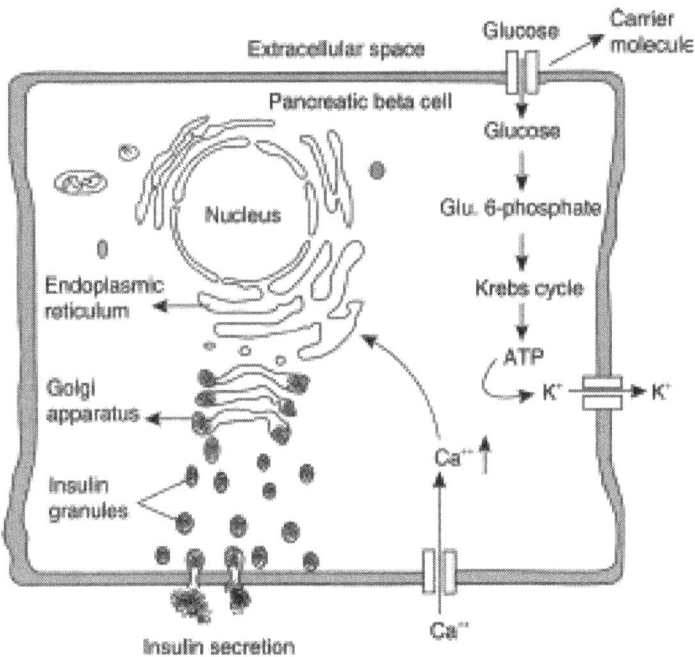

Figure 5.4 Series of events in the biosynthesis and release of insulin from pancreatic beta cell.

5.2.2 Regulation of the Secretion of Insulin

There is a close interrelationship between different types of the cells found in the endocrine pancreas. This allows the direct control of secretion of one hormone by the other. For instance, insulin inhibits the secretion of glucagon, and somatostatin suppresses the release of both insulin and glucagon. The principal stimuli that regulate the secretion of insulin are hyperglycemia (rise in blood sugar level), amino acids in plasma, glucagon, growth hormone, circulating adrenaline, vagal stimulation, and gastrointestinal hormones like pancreozyme and secretin. Increased concentration of glucose in blood has a direct effect on the beta cells of islets of Langerhans, as a result of which, beta cells produce copious amounts of insulin. The insulin in turn causes the excess glucose to be transported into the cells where it can be used for energy, stored as glycogen, or converted into fat.

Thus, there is a feedback mechanism that exists between the blood glucose level and the secretion of insulin by the beta cells of islets of Langerhans for controlling the concentration of glucose in the blood and extracellular fluid. Hyperglycemic condition promotes the secretion of insulin, which then causes increased utilization of glucose and brings the glucose level back to normal. Conversely, when the blood glucose level falls too low, the rate of secretion of insulin secretion decreases, and the glucose is now retained in the body fluid until its level returns to normal.

The control of glucagon secretion is almost exactly opposite that of insulin. The release of glucagon from alpha cells of islets of Langerhans is influenced by such stimuli as hypoglycemia (decrease in blood glucose level), gastrointestinal hormones like gastrin and pancreozyme, high amino acids and low fatty acids in plasma and exercise. The principal stimulus is the fall in blood glucose level as low as 60 mg per 100 ml of blood, in response to which alpha cells in the endocrine pancreas secrete increased quantities of glucagon in the blood.

Glucagon in turn stimulates glycogenolysis and gluconeogenesis in the liver, as a result of which the rapidly increasing blood glucose concentration gets reduced to the normal level of 90 to 100 mg per 100 ml. The regulating mechanism for both hormones is shown in the figure 5.5.

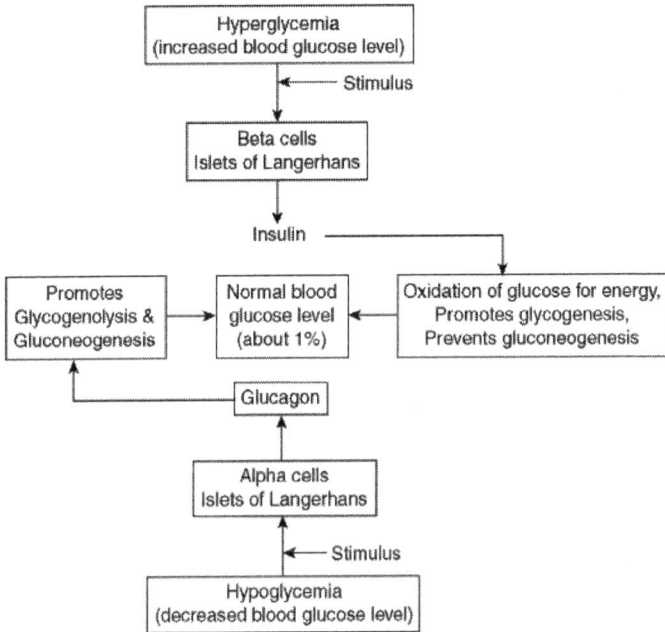

Figure 5.5 Regulation of the secretion of insulin and glucagon.

5.2.3 Biological Action of Insulin

The principal action of the insulin is to control glucose metabolism in the body. It is mainly a sugar-lowering hormone that performs the following biological functions:

i. *Effect of insulin on blood sugar* Insulin promotes transfer of glucose from the extracellular fluid to the interior of the cells by facilitated diffusion and increases the rate of oxidation of glucose. As a result, the concentration of glucose in the blood and extracellular fluid is decreased.

ii. *Effect of insulin on liver and muscles* It accelerates the conversion of glucose to glycogen (glycogenesis) both in the liver and muscles. The hepatocytes of the liver and muscle cells store the glycogen as food reserve, thus helping in the reduction of glucose present in the circulating blood.

iii. *Effect of insulin on protein metabolism* It increases the availability of amino acids in the cells, stimulates transcription of genes to form mRNA, and also promotes ribosomes to synthesize proteins.

iv. *Effect of insulin on fat metabolism* Whenever there is excess glucose in blood, insulin causes some of this glucose to be transported into the fat cells.

v. *Effect of insulin on growth* Since insulin promotes protein synthesis and makes availability of energy from increased oxidation of glucose, it has a promising role in the growth of an organism.

5.2.4 Mechanism of Action of Insulin

As mentioned earlier, insulin is primarily concerned with control of glucose metabolism. To achieve this function, the rate of glucose transport through plasma membrane should be increased. Glucose molecules cannot enter the cells by simple diffusion because of the small size of the pores that are present in the plasma membrane. The transport of glucose across the membrane is facilitated by presence of carrier molecules (*glucose transporter*) in the plasma membrane. Glucose combines with the glucose-carrier molecule and then transported across the cell membrane. Such a process of diffusion that takes the help of a carrier molecule is known as facilitated diffusion. Its mechanism is shown in figure 5.4. The mode of action of insulin on the target cell (signal transduction) for initiation of cellular responses using receptor tyrosine kinases (RTKs) is explained in section 1.14.

5.2.5 Diabetes Mellitus

This condition caused by a deficiency in the secretion or action of insulin, is a relatively common disease ("diabetes mellitus" means "excessive excretion of sweet urine"). Nearly 10% of the Indian population shows some degree of abnormality in glucose metabolism that is indicative of diabetes or a tendency towards the condition. There are two major clinical classes of diabetes mellitus: type I diabetes, or insulin-dependent diabetes mellitus (*IDDM*), and type II diabetes, or non-insulin-dependent diabetes mellitus (*NIDDM*), also called insulin-resistant diabetes. In type I diabetes, the disease begins early in life and quickly becomes severe. This disease responds to insulin injection, because the metabolic defect stems from a paucity of the pancreatic beta cells and a consequent inability to produce sufficient insulin. IDDM requires insulin therapy and careful, lifelong control of the balance between dietary intake and insulin dose. Characteristic symptoms of type I (and type II) diabetes are excessive thirst and frequent urination (*polyuria*), leading to the intake of large volumes of water (*polydipsia*). These symptoms are due to the excretion of large amounts of glucose in the urine, a condition known as *glycosuria*. Type II diabetes is slow to develop (typically in older, obese individuals), and the symptoms are milder and often go unrecognized at first. This is really a group of diseases in which the regulatory activity of insulin is defective. Insulin is produced, but some feature of the insulin-response system is defective. These individuals are insulin-resistant. The connection between type II diabetes and obesity is an active area of research. Individuals with either type of diabetes are unable to take up glucose efficiently from the blood. Another characteristic metabolic change in diabetes is excessive but incomplete oxidation of fatty acids in the liver. The acetyl-CoA produced by oxidation cannot be completely oxidized by the citric acid cycle. Accumulation of acetyl-CoA leads to overproduction of the *ketone bodies* (acetoacetate and hydroxybutyrate). The blood of diabetics also contains acetone, which results from the spontaneous decarboxylation of acetoacetate. Acetone is volatile and is exhaled, and in uncontrolled diabetes, the breath has a characteristic odour sometimes mistaken for ethanol. A diabetic individual who is experiencing mental confusion due to high blood glucose is occasionally misdiagnosed as intoxicated, an error that can be fatal. The overproduction of ketone bodies, called *ketosis*, results in greatly increased concentrations of ketone bodies in the blood (*ketonemia*) and urine (*ketonuria*). The ketone bodies are carboxylic acids, which ionize, releasing protons. In uncontrolled diabetes this acid production can overwhelm the capacity of the blood's bicarbonate buffering system and produce a lowering of blood pH called *acidosis* or, in combination with ketosis, *ketoacidosis*, a potentially life-threatening condition because acidosis often becomes severe enough to cause diabetic coma. Biochemical measurements on blood and urine samples are essential in the diagnosis and treatment of diabetes.

5.2.6 Glucose-tolerance Test

It is a sensitive diagnostic criteria used to define diabetes chemically. The patient fasts overnight, then drinks a test dose of 100 g of glucose dissolved in a glass of water. The blood glucose concentration is measured before the test dose and at 30 min intervals for several hours thereafter. A healthy individual assimilates the glucose readily, the blood glucose rising to no more than about 90 or 100 mg. per 100 ml; little or no glucose appears in the urine. Diabetic individuals assimilate the test dose of glucose poorly; their blood glucose level far exceeds the kidney threshold (about 160 mg per 100 ml), causing glucose to appear in the urine.

5.3 GLUCAGON

It is a protein hormone having 29 amino acid residues in its polypeptide chain with a molecular weight of 3,500 and is secreted by alpha cells of islets of Langerhans. Glucagon is known as *hyperglycemic glycogenolytic factor* (HGF) since it is primarily concerned with increase in blood glucose level.

5.3.1 Biological Action of Glucagon

Many of the functions of glucagon are opposite to those of insulin. Briefly its biological actions may be listed as below:

i. Glucagon stimulates the process of glycogenolysis mainly in liver cells so that there is increase in blood glucose level. The signal transduction mechanism for the action of glucagon on the target cell involves formation of glucagons – receptor complex in the plasma membrane, which in turn activates the enzyme adenyl cyclase to increase the quantity of cyclic AMP inside the cell. Then cAMP mediates cellular response in the form of glycogenolysis.

ii. Similar to insulin, glucagon increases the availability of amino acids and fat and accelerates the rate of conversion of these metabolites into glycogen or glucose, the process is called gluconeogenesis, thereby increasing the concentration of glucose in the circulating blood.

iii. During severe exercise and conditions like starvation, glucagon tends to increase blood glucose level by promoting glycogenolysis and gluconeogenesis and thus protects the body from hypoglycemic attack.

5.3.2 Hypoglycemia

It is defined as a syndrome characterized by low plasma glucose and associated group of symptoms. During the first week after birth, the infant is hypoglycemic, that is, its plasma glucose concentration is less than 25 mg/100 ml of blood. In adults, two groups of symptoms occur, depending upon whether the hypoglycemia is acute or chronic. If the low plasma glucose occurs rapidly, symptoms of sweating, shakiness, trembling, weakness, and anxiety are produced. If the reduction in plasma glucose occurs slowly, headache, irritability, lethargy, and other central nervous system symptoms predominate. Prolonged ingestion of ethanol and certain other drugs, which can damage the liver cells and hamper the supply of liver glycogen for glycogenolysis, may cause fasting hypoglycemia. Adrenocortical insufficiency and hypopituitarism are also associated with hypoglycemia.

5.4 SOMATOSTATIN

It was first discovered in hypothalamic extracts and identified as a hormone that inhibited secretion of growth hormone. Subsequently, somatostatin is secreted by a broad range of tissues, including pancreas (delta cells of islets of Langerhans), intestinal tract and regions of the central nervous system outside the hypothalamus.

5.4.1 Structure and Synthesis of Somatostatin

Two forms of somatostatin are synthesized. They are referred to as SS-14 and SS-28, reflecting their amino acid chain length. Both forms of somatostatin are generated by proteolytic cleavage of prosomatostatin, which itself is derived from preprosomatostatin. Two cysteine residues in SS-14 allow the peptide to form an internal disulphide bond.

The relative amounts of SS-14 versus SS-28 secreted depend upon the tissue. For example, SS-14 is the predominant form produced in the nervous system and apparently the sole form secreted from pancreas, whereas the intestine secretes mostly SS-28. In addition to tissue-specific differences in the secretion of SS-14 and SS-28, the two forms of this hormone can have different biological potencies. SS-28 is roughly ten-fold more potent in inhibition of growth hormone secretion, but less potent that SS-14 in inhibiting glucagon release.

5.4.2 Receptors and Mechanism of Action of Somatostatin

Five somatostatin receptors have been identified and characterized, all of which are members of the G protein-coupled receptor super family. Each of the receptors activates distinct signalling mechanisms within cells, although all inhibit adenyl cyclase. Four of the five receptors do not differentiate SS-14 from SS-28.

5.4.3 Physiological Effects of Somatostatin

Somatostatin acts by both endocrine and paracrine pathways to affect its target cells. A majority of the circulating somatostatin appears to come from the pancreas and gastrointestinal tract. If one had to summarize the effects of somatostatin in one phrase, it would be: "somatostatin inhibits the secretion of many other hormones".

Effects on the pituitary gland   Somatostatin was named for its effect of inhibiting secretion of growth hormone from the pituitary gland. Experimentally, all known stimuli for growth hormone secretion are suppressed by somatostatin administration. Additionally, animals treated with antisera to somatostatin show elevated blood concentrations of growth hormone, as do animals that are genetically engineered to disrupt their somatostatin gene. Ultimately, growth hormone secretion is controlled by the interaction of somatostatin and growth hormone releasing hormone, both of which are secreted by hypothalamic neurons.

Effects on the pancreas   Cells within pancreatic islets secrete insulin, glucagon and somatostatin. Somatostatin appears to act primarily in a paracrine manner to inhibit the secretion of both insulin and glucagon. It also suppresses pancreatic exocrine secretions by inhibiting cholecystokinin-stimulated enzyme secretion and secretin-stimulated bicarbonate secretion.

Effects on the gastrointestinal tract   Somatostatin is secreted by scattered cells in the gastrointestinal (GI) epithelium, and by neurons in the enteric nervous system. It has been shown to inhibit secretion of many of the other GI hormones, including gastrin, cholecystokinin, secretin and vocative intestinal peptide. In addition to the direct effects of inhibiting secretion of other GI hormones, somatostatin has a variety of other inhibitory effects on the GI tract, which may reflect its effects on other hormones, plus some additional direct effects. Somatostatin suppresses secretion of gastric acid and pepsin, lowers the rate of gastric emptying, and reduces smooth muscle contractions and blood flow within the intestine. Collectively, these activities seem to have the overall effect of decreasing the rate of nutrient absorption.

Effects on the nervous system   Somatostatin is often referred to as having neuromodulatory activity within the central nervous system, and appears to have a variety of complex effects on neural transmission. Injection of somatostatin into the brain of rodents leads to such things as increased arousal and decreased sleep, and impairment of some motor responses.

5.4.4 Pharmacological Uses

Somatostatin and its synthetic analogs are used clinically to treat a variety of neoplasms. It is also used to treat gigantism and acromegaly, due to its ability to inhibit growth hormone secretion.

5.5 INSULIN-LIKE GROWTH FACTOR-I (IGF-I)

Growth factors are proteins that bind to receptors on the cell surface, with the primary result of activating cellular proliferation and/or differentiation. Many growth factors are quite versatile, stimulating cellular division in different cell types; while others are specific to a particular cell-type. Table 5.1 shows two types of insulin-like growth factors with their source and activity.

IGF-I (originally called somatomedin C) is a growth factor structurally related to insulin. IGF-I is the primary protein involved in responses of cells to growth hormone (GH): that is, IGF-I is produced in response to GH and then induces subsequent cellular activities, particularly on bone growth. It is the activity of IGF-I in response to GH that gave rise to the term somatomedin. Subsequent studies have demonstrated, however, that IGF-I has autocrine and paracrine activities in addition to the initially observed endocrine activities on bone. The IGF-I receptor, like the insulin receptor, has intrinsic tyrosine kinase activity. Owing to their structural similarities IGF-I can bind to the insulin receptor but does so at a much lower affinity than does insulin itself.

Table 5.1 Source and activity of insulin-like growth factors.

Factor	Principal Source	Primary Activity	Comments
IGF-I	Primarily liver	Promotes proliferation of many cell types	Related to IGF-II and proinsulin, also called Somatomedin C
IGF-II	Variety of cells	Promotes proliferation of many cell types primarily of fetal origin	Related to IGF-I and proinsulin

IGF-1 is a natural anabolic growth factor molecule and is a protein that promotes tissue growth, organ health, and healthy blood sugar levels. Unfortunately, IGF-1 blood levels decrease after the age of thirty, similar to decreases in human growth hormone (hGH). As we age and as lesser growth hormone is released from the pituitary gland, there is a corresponding drop in IGF-1 levels, yet the body's demand for IGF-1 does not decrease. As IGF-1 levels decline, the vitality and physical exuberance of youth recedes, further contributing to the process and experience of aging. Studies have found that balancing hGH and IGF-1 feedback loops can be done to increase the quality of living in a stressful world.

Anticancer therapeutic scientists report that an unlikely molecule has emerged as an attractive target for development of therapeutics aimed at a diverse spectrum of tumours, including some malignancies that are resistant to conventional therapies. Two studies published online in *Cancer Cell* demonstrate that the insulin-like growth factor 1 receptor (IGF-1R) is required for the survival of tumour cells and provide direct evidence that inhibition of IGF-R1 using selective small molecules represents a novel potential anticancer treatment.

Extensive studies have suggested that IGF-1R plays a role in the development of human cancers. IGF-1R is present in a broad range of tumour types including multiple myeloma, lymphoma, leukemia, and breast, lung, prostate, and colon cancers. However, IGF-1R has not been viewed as a likely target for cancer therapeutics because many normal cells also contain the protein. Research scientists from Dana-Farber Cancer Institute in Boston and Novartis Institutes for Biomedical Research, Basel demonstrate that IGF-1R inhibition using a variety of methods had potent antitumour effects against many types of cancer cells grown in the laboratory, including cells that are resistant to conventional cancer therapeutics.

Molecular analyses demonstrated that IGF-1R inhibition impacts multiple intracellular signals related to cell proliferation or tumour development and provides possible mechanisms to explain how IGF-1R inhibition can make tumour cells more sensitive to conventional chemotherapy or other anticancer agents. Perhaps most significantly, IGF-1R suppresses tumour growth, prolongs survival, and enhances the antitumour effect of chemotherapy in clinically relevant mouse models of multiple myeloma and other haematological malignancies. The researchers also identify two small molecules that are selective inhibitors of IGF-1R and are active anticancer agents against tumours that contain IGF-1R. These small molecules represent highly attractive potential therapeutics.

According to a study by Dr. Constantine S. Mitsiades of Dana-Farber, he states, "These results suggest that IGF-1R function is critically required for tumour cell survival, but dispensable for survival of normal cells in adult animals. The preclinical activity of IGF-1R inhibitors against a broad spectrum of tumour cells and, importantly, their ability to sensitize tumour cells to a wide range of anticancer agents, highlight the major role of IGF-1R signaling for human malignant cells, and suggest that the molecular pathway of IGF-1R is an attractive potential target for development of anticancer therapeutics."

5.6 INSULIN-LIKE GROWTH FACTOR-II (IGF-II)

It is almost exclusively expressed in embryonic and neonatal tissues. Following birth, the level of detectable IGF-II protein falls significantly. For this reason IGF-II is thought to be a foetal growth factor. The IGF-II receptor is identical to the mannose-6-phosphate receptor that is responsible for the integration of lysosomal enzymes (which contain mannose-6-phosphate residues) to the lysosomes.

SYNOPSIS

○ Pancreas is a dual-function organ with an exocrine part that releases digestive enzymes while the endocrine portion is made up of islets of Langerhans that secrete two important hormones, insulin and glucagon.

○ The beta cells of islet of Langerhans synthesize the hormone *insulin*. The alpha cells secrete the hormone *glucagon* and delta cells secrete the hormone *somatostatin*.

○ Insulin is a protein hormone with 51 amino acids, which is formed directly by genetic translation via messenger RNA in the form of a precursor called *proinsulin* that undergoes post-translational changes to form insulin.

○ Insulin is principally a sugar-lowering hormone that controls glucose metabolism in the body, by increasing the rate of oxidation of glucose, accelerating the conversion of glucose to glycogen (glycogenesis) both in liver and muscles, and by converting extra blood glucose into fat.

○ Diabetes mellitus is the condition caused by a deficiency in the secretion or action of insulin. The disorder is characterized by excessive thirst, frequent urination (*polyuria*), the intake of large volumes of water (*polydipsia*), and excretion of large amounts of glucose in the urine (*glucosuria*).

○ Glucagon is a protein hormone having 29 amino acid residues secreted by alpha cells of islets of Langerhans. It is known as *hyperglycemic glycogenolytic factor* (HGF) since it is primarily concerned with increase in blood glucose level.

❏ Hypoglycemia is defined as a syndrome characterized by low plasma glucose and associated group of symptoms. It can be acute or chronic showing symptoms like sweating, trembling, weakness, anxiety, headache, irritability, and lethargy.

❏ Somatostatin is a hormone that inhibits secretion of growth hormone. Subsequently, somatostatin is secreted by a broad range of tissues, including pancreas (delta cells of islets of Langerhans), intestinal tract and regions of the central nervous system outside the hypothalamus.

❏ Somatostatin primarily inhibits the secretion of both insulin and glucagon. Pharmacologically, somatostatin and its synthetic analogs are used clinically to treat a variety of neoplasms. It is also used to treat gigantism and acromegaly, due to its ability to inhibit growth hormone secretion.

❏ Growth factors are proteins that bind to receptors on the cell surface, with the primary result of activating cellular proliferation and/or differentiation. Many growth factors are quite versatile, stimulating cellular division in numerous different cell types, while others are specific to a particular cell-type.

❏ IGF-I (originally called somatomedin C) is a growth factor structurally related to insulin. IGF-I is the primary protein involved in responses of cells to growth hormone (GH). IGF-I is produced in response to GH and then induces subsequent cellular activities, particularly on bone growth.

❏ Insulin-like growth factor 1 receptor (IGF-1R) is required for the survival of tumour cells and provides direct evidence that inhibition of IGF-R1 using selective small molecules represents a novel potential anticancer treatment.

❏ Insulin-Like Growth Factor-II (IGF-II) is almost exclusively expressed in embryonic and neonatal tissues and it is responsible for the integration of lysosomal enzymes.

REVIEW QUESTIONS

1. Explain how pancreas acts as a dual-function organ.

2. Describe the structure of endocrine pancreas.

3. Review the effects of insulin on glucose metabolism with mechanism of action.

4. What role does glucagon play in regulation of blood glucose level?

5. How do the functions of insulin differ from those of glucagon?

6. Explain the process of biosynthesis of insulin.

7. What are the effects of the disease, diabetes mellitus, on cellular metabolism, and why do ketosis and acidosis frequently occur?

8. How does somatostatin interact with glucagon and insulin?

9. What is meant by growth factor? What are the points of differences in insulin-like growth factors II and I?

10. Define IGF-1R. Describe its role in cancer treatment.

11. Write short notes on:

    i.   Pancreatic acini
    ii. Islets of Langerhans
    iii. Proinsulin
    iv. Gluconeogenesis
    v.   Glycogenolysis
    vi. Glucosuria

vii. Acidosis and ketosis

viii. Hypoglycemia

ix. Glucose-tolerance test

x. Diabetes mellitus

xi. Somatostatin

xii. Insulin-like growth factors

xiii. Facilitated diffusion

xiv. Diabetic coma

SIX

GASTROINTESTINAL HORMONES

6.1 INTRODUCTION

In mammals, the integration of the activities of various parts of the digestive system in dealing with ingested food is partly endocrine in character. The sustainability of life depends on the continuous supply of food. The digestion of food is the first physiological event in the nutritional process. Food is usually broken down to simpler chemical compounds in the presence of acids, alkalis and various digestive enzymes, which are secreted by glands in the wall of the stomach and intestine as well as the acini of pancreas. Finally the digested food is absorbed through the wall of the intestine.

The secretory and motor activities of the gastrointestinal tract in mammals is not only controlled by the autonomic nervous system but also by various hormones that are secreted by the localized areas of the alimentary canal. More than 15 types of hormone-secreting enteroendocrine cells have been

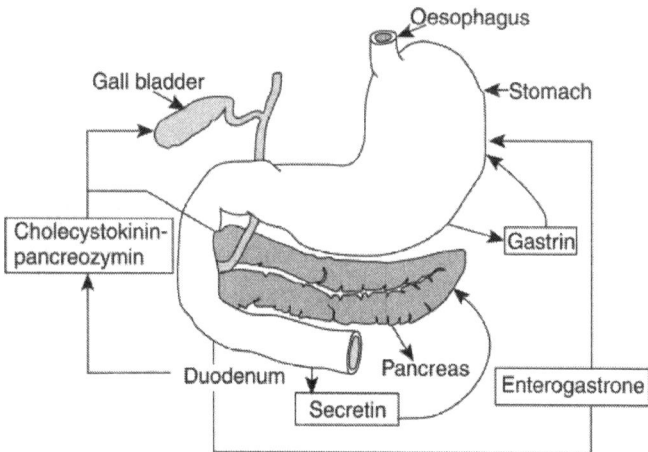

Figure 6.1 Diagrammatic representation of the action of some hormones secreted by gastrointestinal tract with their sites of secretion and action.

identified in the mucosa of the stomach, small intestine and colon. The major hormones and their general sites of formation and regions of action are shown in figure 6.1. The gastrointestinal hormones include *gastrin, secretin, cholecystokinin-pancreozymin, enterogastrone, glucose-dependent insulinotropic polypeptide (GIP), vassoactive intestinal peptide (VIP), motilin.* The release of these hormones is mainly regulated by the presence or absence of particular food in the lumen of the gastrointestinal tract. All the gastrointestinal hormones are protein in nature and play a significant role in regulating the interconversions of nutrients to metabolic substances and their stored forms.

6.2 GASTRIN

The wall of the stomach secretes hydrochloric acid and pepsinogen mainly for the digestion of proteins present in the food. The secretion from the stomach is regulated by both neurogenic and hormonal mechanisms. The food in the stomach can cause local nervous reflexes also called submucosal reflexes, which occur entirely in the wall of stomach itself to cause local secretion of the hormone called *gastrin*. When the protein-rich food (like meat and certain other foods) enters the stomach, the antral portion of the stomach starts secreting the gastrin in the form of a large polypeptide. Then, it passes by way of the blood to the oxyntic cells in the fundus part of the stomach and stimulates them to secrete a strongly acidic gastric juice that helps in the digestion of meats. Stimulation of gastric motility is probably a physiological action as well. The vagus nerve also sensitizes the oxyntic cells to the action of gastrin. The complex reflex arc that is initiated by feeding, initially involves nerve stimulation along cholinergic nerves to the pyloric cells that release the gastrin.

As mentioned earlier, gastrin is a polypeptide hormone that shows differences in molecular structure due to derivatization of single amino acid residues. Its inactive form is called *preprogastrin*, which is processed into fragments of polypeptide chains of various sizes. There are three forms of gastrin called G 34, G 17, and G 14, in which the figures indicate the number of amino acid residues in its polypeptide chain. G 17 is the principal form of gastrin with respect to the gastric acid secretion. G 14 and G 17 have half-lives of 2–3 minutes in the circulation, whereas G 34 has a half-life of 15 minutes. Gastrins are inactivated primarily in the kidney and small intestine.

6.3 SECRETIN

In 1902, Bayliss and Starling first demonstrated that the excitatory effect of duodenal stimulation on pancreatic secretion was due to blood-borne factor. This factor was then identified as *secretin*. When the food (chyme) enters the upper part of the small intestine, it causes a hormone secretin to be released from the glands of the mucosa of the upper portion of the small intestine. Secretin is a linear polypeptide of 27 amino acid residues, having a molecular weight of 3200. The quantity of the secretin released is especially abundant when the chyme is highly acidic. The secretin in turn is absorbed into the blood and then carried to the glandular cells of the pancreas where it causes the cells to secrete large quantities of the fluids containing extra large amount of sodium bicarbonate. This action of secretin on the pancreatic duct cells is mediated via cAMP. The bicarbonate then reacts with the acid of the chyme to neutralize it. Thus, the secretin mechanism is an automatic process to prevent excess acid in the upper part of the small intestine.

The product of protein digestion and acid present in the food that bathes the intestinal mucosa stimulates the upper small intestine to release secretin. This is another example of feedback control. In response to secretin the alkaline pancreatic juice floods into the duodenum, neutralizes the acid from the stomach and thus prevents the release of secretin.

6.4 CHOLECYSTOKININ-PANCREOZYMIN (CCK-PZ)

There is another hormone that is liberated from intestinal mucosa; initially it was called pancreozymin, which promotes secretion of pancreatic enzymes. Also, concentration of the gall bladder to liberate bile is caused by an intestinal hormone, cholecystokinin. It is considered that these activities are due to the same substance and hence considered as a single hormone called CCK-PZ or most commonly CCK. Like gastrin, CCK also exhibits in Prepro-CCK that is processed into many fragments, like CCK 58, CCK 39, CCK 12, CCK 8, and CCK 4, where the figures indicate number of amino acid residues in the polypeptide chain. The half-life of the circulating CCK is about 5 minutes.

The presence of fat in intestine initiates the release of bile from the gall bladder, and it also involves an endocrine reflex due to release of CCK. This hormone has several functions in the body and, apart from its action on gall bladder, endocrine and exocrine pancreas and intestinal motility, it can elicit a sensation of satiety. It was experimentally proved that CCK-injected rats showed reduced feeding. Some of the forms of CCK are also found in the brain where they may be related to production of anxiety and analgesia. In addition to the above-mentioned functions, CCK also inhibits gastric emptying, exerts a trophic effect on pancreas, and may enhance the motility of the small intestine and colon.

It is also evident that along with secretin, it promotes the contraction of the pyloric sphincter and prevents the reflex of duodenum into the stomach. There is a sort of positive feedback mechanism for controlling the secretion of CCK from the small intestine since its secretion is increased when the intestinal mucosa comes in contact with the products of digestion, particularly peptides and amino acids, and also by presence of fatty acids. In response to CCK, bile and pancreatic juice enter the duodenum and digest the protein and fatty food so that more amino acids and fatty acids are formed, which in turn stimulate the intestinal mucosa to secrete more CCK.

6.5 ENTEROGASTRONE

The gastrointestinal hormones with inhibitory effects have also been described. Enterogastrone is one such hormone that is produced in the small intestine in the presence of fatty substances in the duodenum but acts on the stomach by inhibiting its movements and HCl secretion. The ingestion of fatty meals thus retards gastric activity. The action of enterogastrone appears to be principally antisecretory, and since in the presence of enterogastrone the gastric juice produced under the influence of gastric stimulants such as histamine is low in acid content and rich in pepsin, it would appear that enterogastrone preferentially inhibits the acid-secreting parietal cells of the stomach.

6.6 GLUCOSE-DEPENDENT INSULINOTROPIC POLYPEPTIDE (GIP)

It is the hormone secreted by the mucosa of the duodenum and jejunum. It contains 43 amino acid residues. The secretion of GIP is stimulated by the glucose and fat in the duodenum. Previously this hormone was known as gastric inhibitory peptide because in large doses it inhibits the gastric secretion and motility. However, it is now evident that it does not inhibit gastric motility when administered in smaller doses and raises the blood glucose level as seen after a meal. GIP also stimulates β-cells in pancreatic islets for insulin secretion when administered in doses that produce blood levels comparable to those produced by oral glucose. For this reason, it is often called glucose-dependent insulinotropic polypeptide.

6.7 VASSOACTIVE INTESTINAL PEPTIDE (VIP)

It is the hormone present in nerves in the gastrointestinal tract, containing 28 amino acid residues in its polypeptide chain. VIP is also found in blood, in which it has a half-life of 2 minutes. The principal action of VIP in intestine is to promote the intestinal secretion of electrolytes and hence of water. In addition, it also relaxes intestinal smooth muscle, including sphincters, dilates peripheral blood vessels, and inhibits gastric acid secretion.

Other gastrointestinal chemical coordinators have been proposed on the basis of brief experiments, but these appear to have a more questionable existence.

6.8 MOTILIN

It is a polypeptide hormone containing 22 amino acid residues and is secreted by stomach, small intestine and colon. It acts on G protein-coupled receptors on enteric neurons in the duodenum and colon and upon injection it causes contraction of smooth muscles in the stomach and intestine.

6.9 OTHER GASTROINTESTINAL HORMONES

In addition to the above-mentioned hormones, other gastrointestinal hormones are:

i. *Ghrelin* It is the hormone secreted by stomach with secretion decreased by feeding and increased by fasting. It may control the food intake.

ii. *Neurotensin* It is a polypeptide hormone containing 13 amino acid residues that is secreted by neurons and mucosa of the ileum in response to fatty acids. It inhibits gastrointestinal motility and increases ileal blood flow.

iii. *Substance P* It is found in endocrine and nerve cells in the gastrointestinal tract and may enter the circulation. It increases intestinal motility.

iv. *Somatostatin* In addition to hypothalamus and pancreatic islets, gastrointestinal mucosa also secretes this hormone, which exists in tissues in two forms, somatostatin 14 and somatostatin 28. It inhibits gastric acid secretion and motility, pancreatic exocrine secretion, gall bladder contraction, and the absorption of glucose, amino acids and fatty acids.

v. *Glucagon*   The hormone from gastrointestinal tract may be responsible for the hyperglycemia after pancreatectomy.

vi. *Guanylin*   It is a gastrointestinal hormone secreted by the cells of intestinal mucosa and contains 15 amino acid residues in its polypeptide chain that binds guanylyl cyclase. As a consequence, there is increase in intracellular concentration of cGMP and secretion of Cl^- into the intestinal lumen.

SYNOPSIS

❍ Mammalian digestive-control mechanisms involve both nervous and hormonal components.

❍ The gastrointestinal hormones include gastrin, secretin, cholecystokinin-pancreozymin (CCK-PZ), enterogastrone, and vassoactive intestinal peptide (VIP), glucose-dependent insulinotropic polypeptide (GIP) and motilin.

❍ In response to protein-rich food, the antral part of stomach starts secreting the *gastrin,* which passes by way of blood to the oxyntic cells in the fundus part of the stomach and stimulates them to secrete a strongly acidic gastric juice that helps in the digestion of food.

❍ The food (chyme) stimulates the upper part of small intestine to release a hormone known as *secretin* from the intestinal mucosa. *Secretin* is carried to the pancreas where it promotes the cells to secrete large quantities of the fluids containing very large amounts of sodium bicarbonate for neutralizing the acid of the chyme.

❍ *Cholecystokinin-pancreozymin* (CCK-PZ) is another hormone that is liberated from intestinal mucosa, which initiates the release of bile from the gall bladder and promotes secretion of pancreatic enzymes.

❍ *Enterogastrone* is an antisecretory hormone produced in the small intestine in the presence of fatty substances in the duodenum and acts on the stomach by inhibiting its movements and HCl secretion.

❍ *Glucose-dependent insulinotropic polypeptide* (GIP) raises the blood glucose level and also stimulates β-cells in pancreatic islets for insulin secretion.

❍ *Vassoactive intestinal peptide* (VIP) is the hormone that is present in all areas of the gastrointestinal tract and relaxes smooth muscle, inhibits acid secretion, and stimulates pancreatic secretion.

❍ *Motilin* is secreted by the stomach, small intestine and colon that acts on G protein-coupled receptors and causes contraction of smooth muscles in the stomach and intestine.

❍ Other gastrointestinal hormones include *ghrelin, neurotensin, substance P, somatostatin, glucagon,* and *guanylin.*

REVIEW QUESTIONS

1. Discuss the role of the different gastrointestinal hormones in digestion of food.

2. How do gastrin and enterogastrone influence the function of stomach?

3. Explain the role of CCK-PZ and secretin in the regulation of gall bladder and pancreas.

4. In what way does the food in the small intestine get neutralized?

5. Review the types of CCK with their functions.

6. Write short notes on
   i. Vassoactive intestinal peptide
   ii. CCK-PZ

iii. Gastrin

iv. Enterogastrone

v. GIP

vi. Motilin

SEVEN

THE PINEAL BODY

7.1 STRUCTURE OF PINEAL BODY

In man, the pineal body, also called the *epiphysis cerebri,* is relatively a small organ shaped like a pine cone (hence its name). It is located in the geometric centre of the brain directly behind the eyes, below the corpus collosum and posterior to the third ventricle (figure 7.1). The pineal body originates as a sac-like evagination from the dorsal part of the diencephalon. It is reddish-gray in colour and is approximately 10 mm in length in man, whereas in dogs it is only about 1 mm long. It is surrounded by a fine capsule and is composed of epithelial cells arranged to form lobules that are surrounded by fine connective tissue. To observe the pineal, reflect the cerebral hemispheres laterally and look for a small grayish bump in front of the cerebellum.

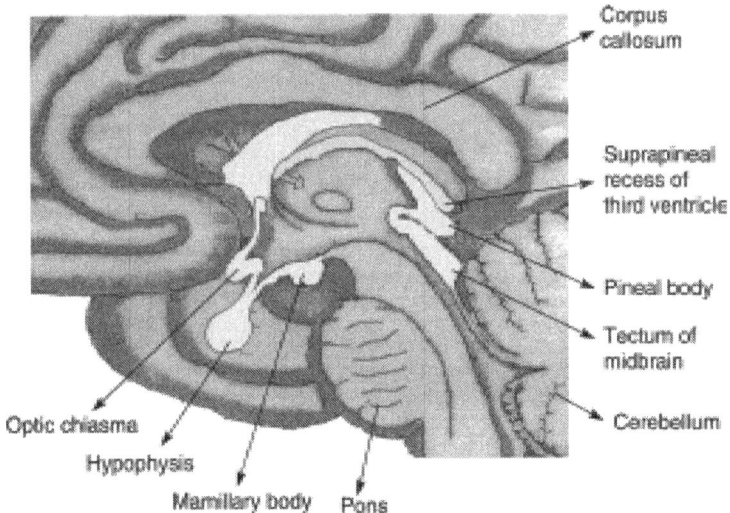

Figure 7.1 The pineal gland in the brain.

Histologically, the pineal is composed of "pinealocytes" and glial cells. Pinealocytes are like neurons. They are stellate (star-shaped) and arranged in clusters. They may have numerous microtubules, extensive smooth-surfaced endoplasmic reticulum and a few small granules. Cells are sensitive to light and innervated by the sympathetic and parasympathetic nervous systems. They receive signals via the retina of the eye. Thus, nerve endings are also seen in the pineal. The nuclei of the pinealocytes are usually pale.

Glial cells are elongated and run between nests of pinealocytes. Often their nuclei are more dense and ovoid. One can differentiate the glial cells from the pinealocytes by their dense nuclei.

With age, the pineal accumulates large calcium-rich particles called "brain sand". This is used to identify the pineal in radiographs and hence serves as a midline landmark.

7.1.1   Pineal body as putative endocrine gland

The pineal gland or epiphysis synthesizes and secretes *melatonin*, a structurally simple hormone that communicates information about environmental lighting to various parts of the body. Ultimately, melatonin has the ability to entrain biological rhythms and has important effects on reproductive function of many animals. The light-transducing ability of the pineal gland has led some to call the pineal the "third eye".

7.2 MELATONIN—SYNTHESIS AND SECRETION

The precursor to melatonin is serotonin, a neurotransmitter that itself is derived from the amino acid tryptophan. Within the pineal gland, serotonin is acetylated and then methylated to yield melatonin.

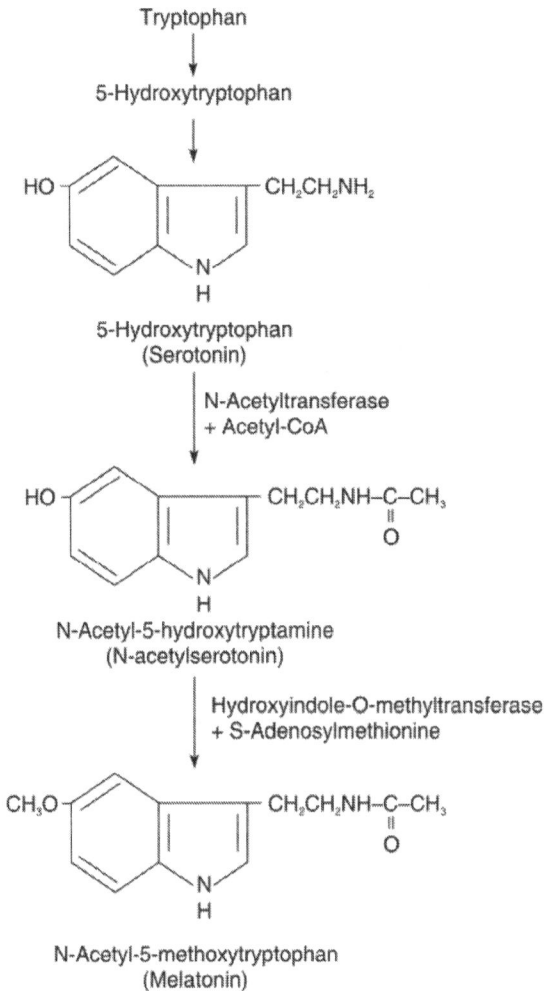

Figure 7.2 Pathway for synthesis of melatonin.

Synthesis and secretion of melatonin (figure 7.2) is dramatically affected by light exposure to the eyes. The fundamental pattern observed is that serum concentrations of melatonin are low during the daylight hours, and increase to a peak during the dark (figure 7.3). The mechanism behind the pattern of secretion during the dark cycle is that activity of the rate-limiting enzyme in melatonin synthesis, serotonin N-acetyltransferase (NAT), is low during daylight and peaks during the dark phase. In some species, circadian changes in NAT activity are tightly correlated with transcription of the NAT messenger RNA, while in other species, post-transcriptional regulation of NAT activity is responsible. Activity of the other enzyme involved in the synthesis of melatonin from serotonin, the methyltransferase, does not show regulation by pattern of light exposure.

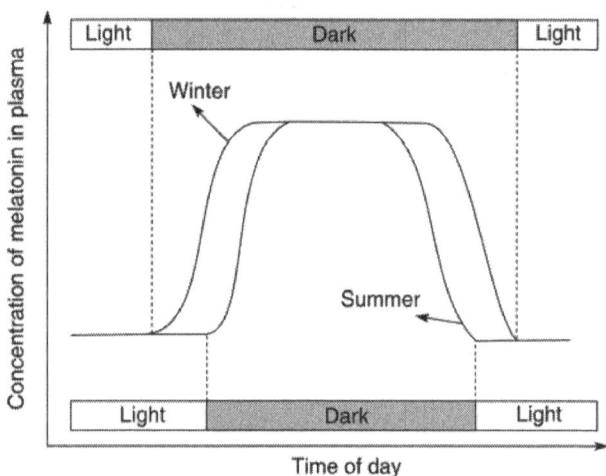

Figure 7.3 Melatonin and photoperiodicity. Release of melatonin is accelerated in dark and thus marks the length of the night while exposure to the light inhibits the release of melatonin. The level of plasma melatonin changes as per the day length.

There is a pathway from the retina to the hypothalamus called the retinohypothalamic tract. It brings information about light and dark cycles to a region of the hypothalamus called the suprachiasmatic nucleus (SCN). From the SCN, nerve impulses travel via the pineal nerve (sympathetic nervous system) to the pineal gland. These impulses inhibit the production of melatonin. When these impulses stop (at night, when light no longer stimulates the hypothalamus), pineal inhibition ceases and melatonin is released. The pineal gland is therefore a photosensitive organ and an important timekeeper for the human body.

Retinal research done with hamsters demonstrates another centre for melatonin production. Located in the retina, this centre implies that the eyes have their own built-in circadian timepiece. This retinal system is distinct from the brain's body clock in the suprachiasmatic nucleus (SCN). Biologists found that they could set and reset the retinal clock even when the SCN was destroyed.

Two melatonin receptors have been identified from mammals (designated Mel 1A and Mel 1B) that are differentially expressed in different tissues and probably participate in implementing differing biological effects. These are G protein-coupled cell surface receptors. The highest density of receptors has been found in the suprachiasmatic nucleus of the hypothalamus, the anterior pituitary (predominantly pars tuberalis) and the retina. Receptors are also found in several other areas of the brain.

7.3 BIOLOGICAL EFFECTS OF MELATONIN

Melatonin has important effects in integrating photoperiod and affecting circadian rhythms. Consequently, it has been reported to have significant effects on reproduction, sleep-wake cycles and other phenomena showing circadian rhythm.

7.3.1 Effects on Reproductive Function

Seasonal changes in day length have profound effects on reproduction in many species, and melatonin is a key player in controlling such events. In temperate climates, animals like hamsters, horses and sheep have distinct breeding season. During the non-breeding season, the gonads become inactive (e.g. males fail to produce sperm in any number), but as the breeding season approaches, the gonads must be rejuvenated.

The effect of melatonin on reproductive systems can be summarized by saying that it is anti-gonadotropic. Melatonin inhibits the secretion of the gonadotropic hormones, luteinizing hormone (LH) and follicle-stimulating hormone (FSH), from the anterior pituitary. It therefore prevents onset of puberty before the appropriate age. Children who experience precocious puberty may have a brain tumour that destroyed the pineal gland.

Seasonal breeders are regulated by the length of day (signaled by sunlight). Long days inhibit melatonin and stimulate gonadal activity. Shorter days stimulate melatonin, which causes gonadal atrophy (via the hypothalamus and pituitary). One practical application of melatonin's role in controlling seasonal reproduction is found in its use to artificially manipulate cycles in seasonal breeders. For example, sheep that normally breed only once per year can be induced to have two breeding seasons by treatment with melatonin.

7.3.2 Effects on Sleep and Activity

Melatonin is probably not a major regulator of normal sleep patterns, but undoubtedly has some effect. One topic that has garnered a large amount of interest is using melatonin alone, or in combination with phototherapy, to treat sleep disorders. There is some indication that melatonin levels are lower in elderly insomniacs relative to age-matched non-insomniacs, and melatonin therapy in such cases appears modestly beneficial in correcting the problem.

Another sleep disorder is seen in shift workers, who often find it difficult to adjust to working at night and sleeping during the day. The utility of melatonin therapy to alleviate this problem is equivocal and appears not to be as effective as phototherapy. Still another condition involving disruption of circadian rhythms is jet lag. In this case, it has repeatedly been demonstrated that taking melatonin close to the target bedtime of the destination can alleviate symptoms; it has the greatest beneficial effect when jet lag is predicted to be worst (e.g. crossing many time zones).

In various species of mammals including humans, administration of melatonin has been shown to decrease motor activity, induce fatigue and lower body temperature, particularly at high doses. The effect on body temperature may play a significant role in melatonin's ability to entrain sleep-wake cycles, as in patients with jet lag.

7.3.3 Other Effects of Melatonin

In addition to above-mentioned effects of melatonin on biological systems, melatonin may involve in distribution of melanin pigment in amphibians and migration of birds. One of the first experiments conducted to elucidate the function of the pineal, extracts of pineal glands from cattle were added to water containing tadpoles. Interestingly, the tadpoles responded by becoming very light in colour or almost transparent due to alterations in melanin pigment distribution. Although such cutaneous effects of melatonin are seen in a variety of lower vertebrates, the hormone does not have such effects in mammals or birds.

Studies done mostly with birds strongly suggest that the pineal gland is a centre for navigation. Scientists believe that the pineal body is a magnetoreceptor, capable of monitoring magnetic fields, and helping to align the body in space. Changing the direction of magnetic fields around the heads of birds alters their ability to orient.

SYNOPSIS

- ❍ The pineal body or *epiphysis cerebri* is located in the geometric centre of the brain directly behind the eyes, below the corpus callosum and posterior to the third ventricle. Due to its light-transducing ability the pineal gland is called "third eye".

- ❍ Histologically, the pineal is composed of "pinealocytes" and glial cells. Pinealocytes are sensitive to light and innervated by the sympathetic and parasympathetic nervous systems. They receive signals via the retina of the eye. Glial cells are elongated with dense nuclei and run between nests of pinealocytes.

- ❍ The pineal gland or epiphysis synthesizes and secretes *melatonin,* from the precursor serotonin, which is a neurotransmitter derived from the amino acid tryptophan.

- ❍ Synthesis and secretion of melatonin is low during the daylight hours and increase to a peak during the dark.

- ❍ There are two melatonin receptors found in mammals designated Mel 1A and Mel 1B, which are differentially expressed in different tissues and participate in implementing differing biological effects.

- Melatonin shows *anti-gonadotropic effect* as it inhibits the secretion of luteinizing hormone (LH) and follicle-stimulating hormone (FSH) from the anterior pituitary and hence prevents onset of puberty before the appropriate age.

- The normal sleep patterns may be regulated by melatonin. Administration of melatonin in mammals is associated with decrease in motor activity, fatigue and lower body temperature.

- Melatonin is also involved in distribution of melanin pigment in tadpoles and migration of birds.

REVIEW QUESTIONS

1. Review the position and structure of pineal gland.

2. Explain the synthetic pathway of melatonin.

3. Describe various biological effects of melatonin.

4. How does melatonin interact with gonads?

5. Discuss the role of melatonin in phototherapy.

6. Write short notes on
   i. Circadian rhythm
   ii. Suprachiasmatic nucleus
   iii. Melatonin receptors
   iv. Serotonin N-acetyltransferase
   v. Third eye.

EIGHT

HORMONES AND
REPRODUCTION

8.1 INTRODUCTION

Reproduction is an essential process for the perpetuation of the species since it involves the formation of the new organisms of the same kind as the parents either by asexual or sexual method. In lower forms of life like *Amoeba*, the reproduction may be an *asexual* process in which the genetic material is equally distributed between two daughter cells simply by mitosis. In multicellular organisms such as *Hydra*, jellyfishes, corals, etc. budding is an additional mode of asexual reproduction while in *Planaria*, starfishes, etc. the small segment broken off from the parent body can produce a new individual. The animals produced as a result of asexual reproduction are at a disadvantage from the point of view of their genetic variability and hence they have lesser chances of evolutionary adaptability.

In *sexual* reproduction, the offsprings are produced as a result of the meeting of two parents or the union of cells from each parent. Though some of the lower animals are hermaphrodite or bisexual, their meeting is essential for the successful cross-fertilization. In higher vertebrates, sexual reproduction involves the production of gametes from gonads by reduction division or meiosis, union of opposite partners for subsequent fusion of sperm and ovum, and formation of the zygote that develops into the offspring similar in appearance and character to that of the parents. The offsprings thus produced as a result of sexual reproduction have genetic variations and hence play significant role in organic evolution.

The patterns of sexual reproduction in vertebrates show several major differences with considerable effects on the endocrine control over reproduction. It is related with the manner by which fertilization is accomplished and the site where the embryo differentiates and grows. In mammals, Prototheria or monotremes are reptile-like and oviparous while in the Theria, the marsupials give birth to immature young ones, which complete their development in the abdominal pouch of the female. The highly evolved placental mammals are either continuous or seasonal breeders and their females undergo oestrous or menstrual cycle. After attaining puberty, the events like mating, fertilization, pregnancy, parturition, and lactation are all under the hormonal influence. Thus, sexual reproduction is a complex process that is based on the social, morphological and physiological coordination.

8.2   DISORDERS OF SEXUAL DIFFERENCES IN HUMANS

Reproduction involves the transmission of genetic material in the form of genes that are present on the chromosomes in specialized cells called gametes. The male gametes are spermatozoa formed in the testes and female gametes are ova produced in ovaries during the process of gametogenesis called spermatogenesis and oogenesis respectively. In human beings there are several abnormalities in sexual development, which could be caused by genetic or hormonal abnormalities, as well as by other nonspecific teratogenic influences.

8.2.1   Chromosomal Abnormalities

An established defect in gametogenesis is *meiotic nondisjunction*, a phenomenon in which a pair of chromosomes fails to separate, as a result both go to one of the daughter cells during meiosis. Four of the abnormal zygotes that can form as a consequence of meiotic nondisjunction of one of the X chromosomes during oogenesis are shown in figure 8.1. A zygote with only one X chromosome and no second sex chromosome (denoted as XO) develops into sex chromatin-negative female with *Turner's syndrome* or *gonadial dysgenesis,* in which the genotype is 44A+XO with the total number of chromosomes 45. Such individual is characterized by arrested genital development in juvenile state, slightly abnormal body structure, other congenital abnormalities and no maturation at puberty. Since a Y chromosome is male determining, person with at least one Y chromosome develops as male. A male with an extra X chromosome having the genotype 44A + XXY and total number of chromosomes 47, develops as sex chromatin-positive *Klinefelter's syndrome,* which is characterized by mental retardation, under development of genitalia, and presence of feminine physical characteristics (such as breast enlargement). Since it has abnormal seminiferous tubules and the process of spermatogenesis is drastically affected, the syndrome is also known as *seminiferous tubule dysgenesis.* The superfemale with XXX combination of sex chromosomes is only second in frequency to the XXY pattern and may be even more common in the general population, since it is not seen to be associated with any characteristic abnormalities. The individual having the genotype 44A + YO with total number of chromosomes 45 is probably lethal and the fetus dies in utero.

Chromosomal abnormalities can be more complex if nondisjunction occurs in second meiotic division during gametogenesis, which is actually a mitotic division. The faulty mitotic division with nondisjunction in early zygote results in production of a *mosaic,* an individual with two or more population of cells with different chromosome complements. The example of such disorder is a *true hermaphroditism* in which the person has both ovaries and testes and the condition is probably due to XX/XY mosaicism.

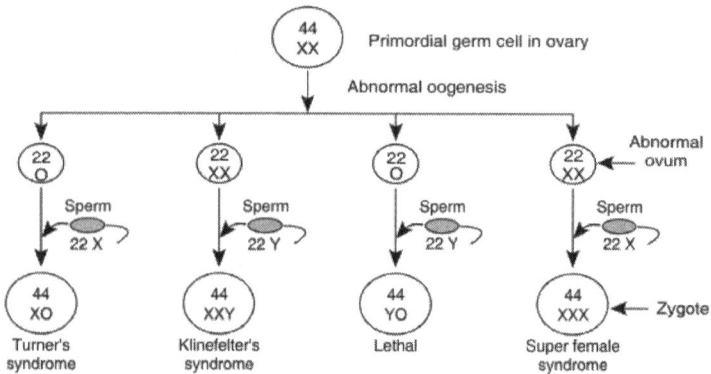

Figure 8.1 The result of four possible defects induced by meiotic nondisjunction.

In addition to nondisjunction of sex chromosome, several different autosomal nondisjunction can also occur to produce abnormalities. One of such disorders is *21 Trisomy* produced as a result of nondisjunction of autosome number 21. Even though chromosome 21 is smallest human chromosome, containing 1,500 genes, the presence of an extra copy of this genetic material has serious consequences producing the condition called *Down's syndrome* or *Mongolism*. Persons with such abnormality are usually mentally retarded, and have alteration in some body features, increased susceptibility to infectious diseases, a greatly increased risk of developing leukemia, and early onset of Alzheimer's disease. In most instances, non disjunction occurs in the ovary rather than the testis and incidence of Down's syndrome increases with advancing age of the mother.

There are several other chromosomal abnormalities and numerous diseases due to the defects in a single base pair in DNA or a gene that can be clinically diagnosed in utero by amniocentesis or biopsy of chorionic villi.

8.2.2 Hormonal Abnormalities

The sex hormones specifically androgens play indispensable role in development of genitalia. The male external genitalia develop normally in genetic male in response to androgen secreted by the embryonic testes. Sometimes the male genital development may also occur in genetic female exposed to androgens from some other source during the eight to the thirteenth weeks of gestation. The condition in which the person suffering from such abnormality is known as *female pseudohermaphroditism*. A pseudohermaphrodite is an individual with the genetic constitution and gonads of one sex and genitalia of the other. After the thirteenth week, the genitalia are fully formed, but exposure to androgens can cause hypertrophy of the clitoris. The condition may be due to congenital virilizing adrenal hyperplasia, or it may be caused by administration of androgens to the mother. Conversely, *male pseudohermaphroditism* is the phenomenon occurring in the genetic male whose testes are defective and has developed external genitalia of female.

Androgen resistance is another cause of male pseudohermaphroditism, in which, as a result of various congenital abnormalities, male hormones cannot exert their full effects on the tissues. *5α-reductase deficiency* is one form of the androgen resistance, in which the gene responsible for formation of this enzyme is mutated. 5α-reductase is necessary for the formation of dihydrotestosterone (DHT-one of the active forms of male sex hormones) from testosterone. Congenital 5α-reductase deficiency leads to an interesting form of male pseudohermaphroditism. Individuals with this defect are born with male genitalia including testes, but they have female external genitalia and are usually grown as girls. Their circulating LH and testosterone levels are increased when they attain puberty. As result, they develop masculine body and male libido. In such circumstances, they usually change their gender identities and become boys.

Mutation in the androgen receptor gene forms the basis for the other form of androgen resistance. The resulting defects in receptor function range from minor to severe. Infertility with or without gynecomastia is the result of mild defect but when there is complete loss of receptor function, the condition may lead to *testicular feminizing syndrome* or *complete androgen receptor syndrome*. It is characterized by presence of female external genitalia but absence of female internal genitalia and hence vagina ends blindly. Individuals with such syndromes also develop enlarged breasts at puberty and usually are considered to be normal women until they are diagnosed for lack of menstruation. The congenital deficiency of *17α-hydroxylase* (the enzyme that catalyzes the formation of testicular and adrenal androgens from the precursor steroid molecule-pregnenolone) can also be one of the causes of male pseudohermaphroditism.

8.3 PUBERTY

In all mammals there is a period in which the gonads of both the sexes are activated by the gonadotropins secreted by anterior pituitary to bring about the final maturation of the reproductive system. This period of final maturation is called *adolescence* or *puberty*. It is defined as the period when the endocrine and gametogenic functions of the gonads have first developed to the point where the reproduction is possible. In girls, the sign is the development of the breast. The event is called *thelarche* that is followed by *pubarche*, the development of axillary and pubic hair, and then *menarche*, the first menstruation period. In the initial periods, there is no ovulation from the ovary and regular ovulation occurs about a year later. The age at the time of puberty is variable. As per the recent study, in most of the countries, puberty generally occurs between the age of 8 to 13 in girls and 9 to 14 in boys. Puberty is also marked by the increase in the secretion of adrenal androgens, the most important of which is dihydroepiandrosterone. The onset of this increase is called *adrenarche*. Usually it occurs at age 8–10 years in girls and age 10–12 years in boys.

8.3.1 Gonadotropins and Puberty

Gonadotropins are secreted by anterior pituitary gland in the form of FSH and LH, while their secretion is under the control of hypothalamic factor that are collectively called GnRH (see chapter 2.2.14 and 2.3). Gonadotropins, FSH and LH, can stimulate the gonads of children; their pituitaries contain gonadotropins and their hypothalami contain GnRH. However their gonadotropins are not secreted. It is experimentally proved that in immature monkeys, normal menstrual cycles can be brought by pulsatile injection of GnRH, and they persist as long as the pulsatile injection is continued. Thus it indicates clearly that pulsatile secretion of GnRH is responsible for onset of puberty.

8.3.2 Precocious Puberty

It is an early development of secondary sexual characteristics without spermatogenesis or oogenesis in immature males and females when they are abnormally exposed to androgen and oestrogen respectively. Such syndrome is called *precocious pseudopuberty* and it is distinguished from *true precocious puberty*, the condition that appears early but otherwise with normal pubertal pattern of gonadotropin secretion from the pituitary. There is another disorder called *constitutional precocious puberty* in which no cause can be determined and that is more common in girls than in boys. In one example of this disorder a 3-year-old girl developed pubic hair and started to menstruate at the age of 17 months. The condition can be associated with the tumors or infections that cause secondary damage to the hypothalamus and interrupt the normal pathway of secretion of GnRH.

8.3.3 Delayed Puberty and Absence of Puberty

The age at which adolescent changes occur is so wide that one cannot say that puberty is delayed pathologically until the menarche has failed to occur by the age of 17 or testicular development by the age of 20. The maturation of a girl or a boy may be due to panhypopituitarism that is associated with dwarfism and other endocrine abnormalities like that in Turner's syndrome or the patient with gonadal dysgenesis. There may be the case of delayed puberty even though the gonads are present and other endocrine functions are normal, such clinical condition is known as *eunuchoidism* in male and *primary amenorrhea* in female.

8.4 SEX CHROMATIN

It is a mass of condensed chromatin, also known as Barr body usually found lying against the inner surface of the nuclear membrane of the somatic cells as shown in figures 8.2 and 8.3. The sex of an individual can also be distinguished by identification of sex chromatin. Since Barr body is derived from X chromosome, it is present only in genetic female. Sex chromatin is found in the polymorphonuclear leucocytes, epithelial cell lining of the buccal mucosa and vaginal smear, neurons, etc., of the female mammals except rodents. The detection of the Barr body forms a diagnostic tool for normal genetic female (44A + XX) and sex chromatin-positive male or Klinefelter's syndrome (44A + XXY).

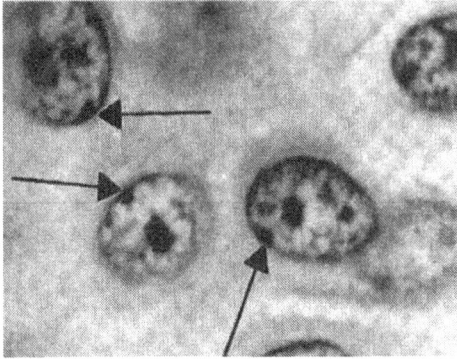

Figure 8.2 Sex chromatin in epithelial cells of a woman indicated by arrows.

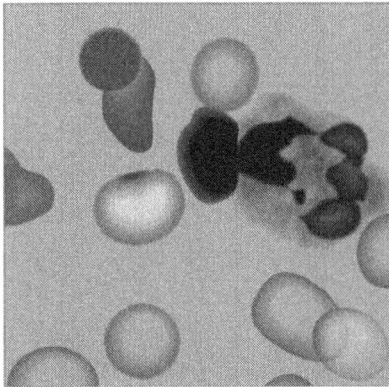

Figure 8.3 Sex chromatin in polymorphonuclear neutrophil of a woman.

8.5 SEXUAL FUNCTIONS OF MALE AND FEMALE

Both male and female play equal roles in initiating the process of reproduction and also in determining hereditary characteristics of the baby. After attending puberty, the male provides the sperm and the female the ovum. A single sperm fertilizes a single ovum to form a zygote that grows into an embryo, and then into a foetus, and finally into a newborn baby. The pregnancy, parturition and lactation are integral processes of reproduction. The endocrine system and its indispensable role in the sexual functions of male and female are discussed in the present chapter.

8.6 MALE REPRODUCTIVE SYSTEM

Figure 8.4 illustrates the various parts of the male reproductive system in man, the most important of which are the testes that are made up of loops of convoluted seminiferous tubules where the spermatozoa are formed from primordial germ cells (spermatogonia) in the process of spermatogenesis.

The testes are reproductive glands of the male that are enclosed in a pouch of deeply pigmented skin called scrotum, which mainly protect the sperms from adverse impact of temperature. Each testis is surrounded by three layers of tissues namely, tunica vaginalis, tunica albuginea, and tunica vasculosa. The sperms formed in convoluted seminiferous tubules of the testes are passed to the head of epididymis through the straight seminiferous tubules and vasa efferentia. From there, the spermatozoa pass through the tail of the epididymis into vas deferens, which at first is a coiled tube then straightens out and leaves the testis and the scrotum

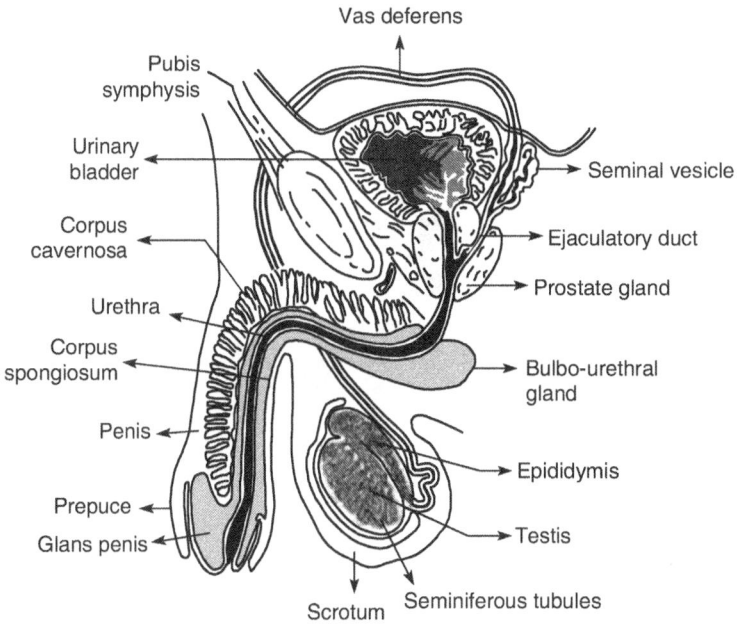

Figure 8.4 Male reproductive system.

enclosed within the spermatic cord (figure 8.5). Both spermatic cords, one leading from each testis, pass through inguinal canal and enter through the ejaculatory ducts. On the way, they receive the ducts from seminal vesicles and then pass through prostate gland and ultimately merge with the internal urethra. In addition, a bulbourethral or Cowper's gland also opens into the urethra. At the time of ejaculation, in addition to sperms, mucus from seminal vesicles and a milky serous fluid from prostate (collectively all these fluids are called semen) are expelled from urethra that finally opens to the exterior through penis. The penis in turn has two important parts for performance of the sexual act. These are the glans, which is sensitive portion that elicits sexual excitement, and the erectile tissues viz. corpora cavernosa and corpora spongiosum, which

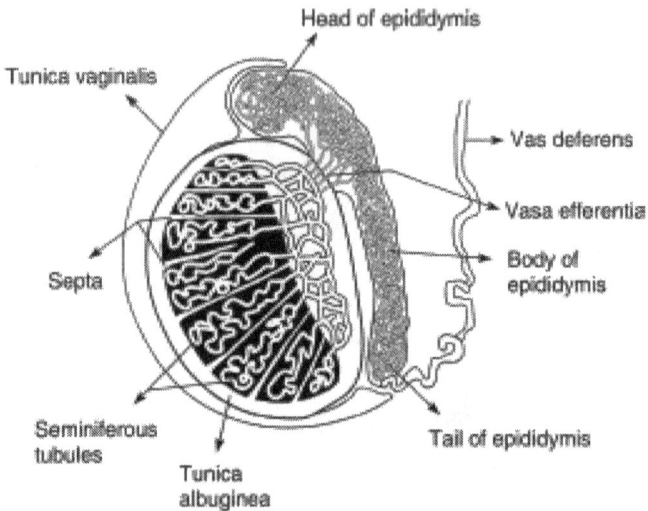

Figure 8.5 A longitudinal section of testis showing arrangement of ducts.

surround the urethra and cause erection. When the degree of sexual stimulation has reached a critical level during coitus, sympathetic nerves are stimulated causing ejaculation of semen in the female vagina.

8.6.1 Spermatogenesis

It is the process of formation of spermatozoa from the primordial germ cells of the testes. The testes are composed of thousands of seminiferous tubules, which are lined by germinal epithelium and it is from these that all the spermatozoa are formed in the process of spermatogenesis. A cross section of a typical seminiferous tubule is shown in figure 8.6. Spermatogonium derived from germinal epithelium is the earliest form of cell in the development of spermatozoa. Spermatogonium mitotically divides to form diploid primary spermatocytes. Each primary spermatocyte then undergoes meiotic division to form two haploid secondary spermatocytes, which are further processed through equatorial division to form spermatids. Thus each diploid primary spermatocyte produces four haploid, nonmotile spermatids that are finally developed into motile sperms by the process called spermiogenesis. The ripening of sperms takes place while they are deriving nutrition from the cells of Sertoli within the seminiferous tubules. Spermatozoa leaving the testes are not fully mobile. They continue their maturation and acquire motility during their passage through epididymis. The whole process of spermatogenesis takes about 74 days in men. Scattered in between the seminiferous tubules are nests of cells containing lipid granules called interstitial cells of Leydig that form the endocrine part of the testis and secrete the male sex hormone testosterone into the bloodstream.

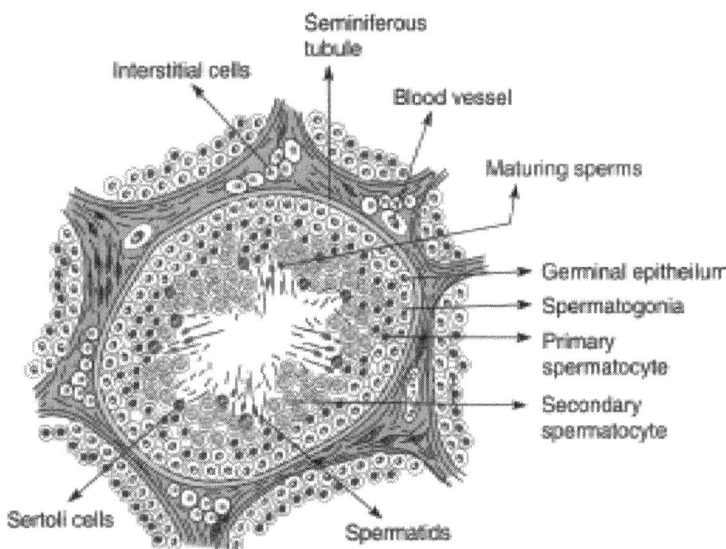

Figure 8.6   A cross section of seminiferous tubule showing different developing stages of the spermatozoa.

Thus, the testes of male produce billions of sperms or spermatozoa, structurally each of which is a single motile cell principally made up of head and tail as shown in the figure 8.7. The head is composed mainly of the nucleus containing a haploid set of chromosomes (22 A + X or 22A + Y). On the top, it has a cap like structure called acrosome, a lysosome-like organelle rich in hydrolytic enzymes including hyaluronidase, which helps the sperm to enter the ovum during fertilization. Proximally the tail consists of middle piece that is marked by the presence of mitochondria and centrioles. Most of the cytoplasm of the sperm is aggregated into the tail that is covered by a long extension of the cellular membrane

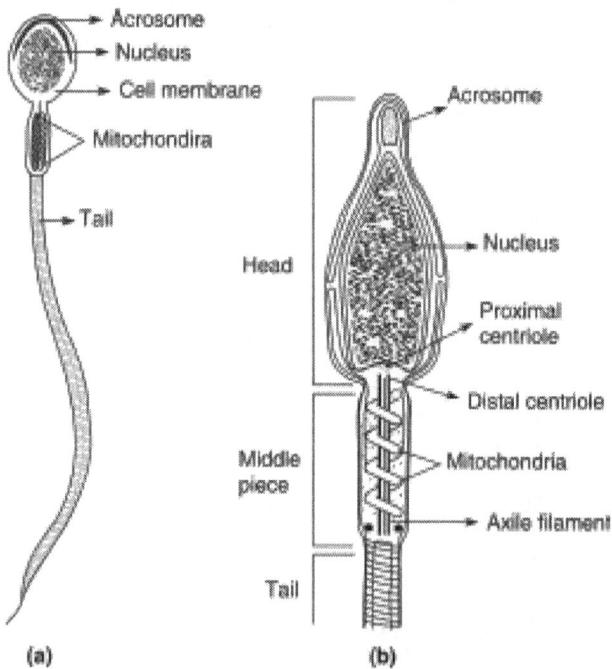

Figure 8.7 Structure of mammalian sperm under electron microscope. (a) complete sperm, (b) anterior end magnified.

enclosing the axial filament. The tail provides motility to the sperm. Once ejaculated into female, the flagellar movement of the tail propels the sperm forward (progressive motility) in the genital tract towards the ovum.

8.7 HORMONAL CONTROL OF TESTES

The process of sperm formation in testes is initiated by the increase in gonadotropic hormones (i) follicle-stimulating hormone (FSH) and (ii) luteinizing hormone (LH). As stated in section 2.3, release of these hormones from the anterior pituitary gland is regulated by the gonadotropin-releasing hormone (GnRH) released by hypothalamus. FSH stimulates the testes for gametogenic function by promoting proliferation of the cells in the germinal epithelium. It accelerates the formation of secondary spermatocytes from primary spermatocytes. LH or interstitial cell stimulating hormone (ICSH) stimulates Leydig cells in the testes to secrete the male sex hormone, testosterone, that plays an indispensable role in the development of secondary sexual characters. The interrelationship between hypothalamus, anterior pituitary and testes is summarized in figure 8.8. Testosterone reduces plasma ICSH, but it has no effect on FSH. Based on the observations, it seems that there is increased level of plasma FSH in patients who have atrophy of the seminiferous tubules but normal levels of testosterone and ICSH. Recently, the role of a factor called *inhibin* is being researched in this connection. It has been an established fact now that there are two inhibins present in the testes and antral fluid of the ovarian follicle. Biochemically, inhibins are formed from three polypeptide subunits (α, β_A, and β_B).

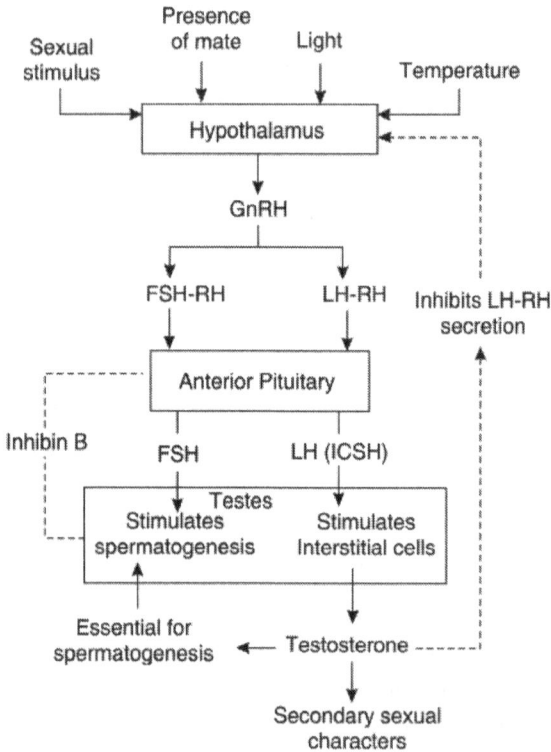

Figure 8.8 Interrelationship between hypothalamus, anterior pituitary and testes.

A glycosylated α subunit has a molecular weight of 18,000 and the two nonglycosylated β subunits, β_A, and β_B, each have a molecular weight of 14,000. The α subunit is linked to β_A subunit by disulphide bonds to form a heterodimer called inhibin A, while inhibin B is formed by the combination of α subunit with β_B. Both the forms of inhibin ($\alpha\beta_A$ and $\alpha\beta_B$) directly act on the pituitary and suppress the secretion of FSH, though it now appears that it is inhibin B that is the FSH-regulating inhibin in men and women. The Sertoli cells in testes and granulose cells in ovary are responsible for synthesis of inhibins.

8.8 CHEMISTRY AND BIOSYNTHESIS OF TESTOSTERONE

Testosterone is the principal hormone of the testes. Chemically, it is a C_{19} steroid with an –OH group in the 17 position (figure 8.9). Testosterone is synthesized from cholesterol in the Leydig cells and is also formed from androstenedione secreted by the adrenal cortex as shown in the biosynthetic pathways in the figure 4.2. In all endocrine glands, the biosynthetic pathways for synthesis of steroid hormones are similar, the organs differing only in the enzyme system they contain. In the Leydig cells, 11- and 21-hydroxylase found in the adrenal cortex are absent,

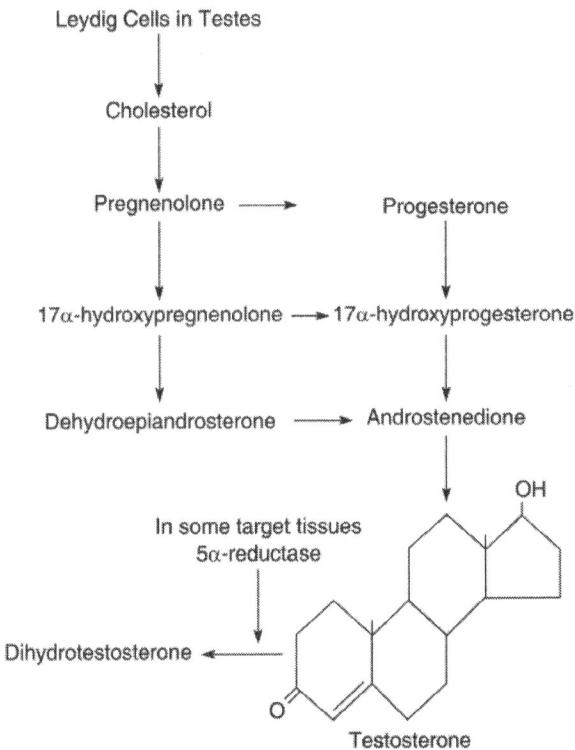

Figure 8.9 Biosynthetic pathway of testosterone.

but 17α-hydroxylase is present. Therefore, pregnenolone is hydroxylated in the 17 position and then subjected to side chain cleavage to form dehydroepiandrosterone, which is also formed via progesterone and 17-hydroxyprogesterone, but this pathway is less prominent in humans. Testosterone is then formed from dehydroepiandrosterone and androstenedione. In some target cells, testosterone is converted to dihydrotestosterone (DHT) by the action of the enzyme 5α-reductase. *Testosterones, DHT,* and *androstenedione* are usually referred to as *androgens*—all are steroid hormones that are secreted by the testes.

Luteinizing hormone influences the secretion of testosterone, and the mechanism by which LH stimulates the Leydig cells involves increased formation of cAMP. Cyclic AMP promotes the formation of cholesterol from cholesteryl esters and conversion of cholesterol to pregnenolone through the activation of protein kinase A.

8.8.1 Functions of Testosterone

Once the testosterone is formed in the Leydig cells, capillaries and veins carry it to the periphery, or it traverses testicular myoid cells and enters the seminiferous tubules. Testosterone performs the following functions:

Effect on spermatogenesis   It causes testes to enlarge. Also, it must be present along with the follicle-stimulating hormone for completion of spermatogenesis. FSH is responsible for activation of seminiferous tubules, resulting in production of sperms as well as the conversion of testosterone to estradiol. In the seminiferous tubules, testosterone stimulates primary spermatocytes to form secondary spermatocytes and finally young spermatids.

Effect on male sex organs Once the foetus destined to be a male starts developing in the mothers uterus, within a few weeks of development, its testes begin to secrete testosterone under the impact of the placental hormone called chorionic gonadotropin. This testosterone helps the foetus to develop male sexual organs. The formation of a penis, a scrotum, a prostate, the seminal vesicles, the vas deferens, and other male sex organs are accelerated under the influence of foetal testosterone.

Since there is loss of connection with placenta immediately after the birth of the child, the stimulatory effect of chorionic gonadotropin is removed and as a result, the testes stop secreting testosterone. As a consequence, the sexual characters cease developing from birth until the onset of puberty. At puberty testes resume the function of testosterone secretion and under its impact there is about tenfold increase in the size of male sex organs including testes, scrotum, and penis.

Effect on secondary sexual characters Testosterone not only affects the growth of male sex organs but also exerts other widespread changes throughout the body known as secondary sexual characters including hair distribution, masculine figure, deeper voice, etc. At puberty, testosterone gives the adult male his distinctive characteristics. The male secondary sexual characters are summarized in Table 8.1. If testes are removed, before the puberty (process is called *castration*), the boy will be unable to develop the secondary sexual characters.

Anabolic effects Androgens is the collective term used for describing the hormones secreted by the testes including testosterone. Androgens are protein

Table 8.1 Male secondary sexual characteristics formed at puberty in boys.

External genitalia	Increase in the length and width of the penis, scrotum becomes pigmented.
Internal genitalia	Seminal vesicles, prostate and Cowper's glands enlarge and begin to secrete.
Hair growth	Mustache and beard appear, decrease in scalp hair, pubic and axillary hair grow, more dense hair also appears on chest and general body surface.
Voice	Size of the larynx increases, length and thickness of vocal cords increases, pitch of sound is lowered and hence voice becomes deeper.
Mental	Aggressiveness, active attitude and interest in opposite sex.
Skin	Sebaceous gland secretion thickens and increases and leads to acne on face.
Configuration of body	Broadening of shoulders and enlargement of muscles.

anabolic hormones. They have direct stimulating effect on DNA or RNA polymerase; as a result transcription of DNA for synthesis of mRNA is accelerated and the rate of protein synthesis is increased. Thus androgens increase the synthesis and decrease the breakdown of protein, leading to nitrogen retention and an increase in the rate of growth. Secondary to this effect, androgens also promote retention of sodium, potassium, calcium, phosphate, and water. Administration of testosterone in the body causes significant anabolic effects, which are also masculinizing and increase libido in addition to augmentation of BMR.

8.9 TESTICULAR ABNORMALITIES

Apart from the defects mentioned in 8.2.2 and the sexually transmitted diseases like syphilis, gonorrheoa, urethritis and tumours or other infections in accessory sexual organs, some of the common testicular defects are discussed in the following sections.

8.9.1 Undescended Testes (Cryptorchidism)

During the development of the foetus, the testes are abdominal and normally migrate to the scrotum. In cryptorchidism, either one or both testes may be arrested at any point along the tract marked by the testicular artery and vein from their origin near the kidneys down through the inguinal canal into scrotum. Thus one or both testes remain in abdominal cavity or inguinal canal. The treatment with gonadotropin can correct the disorder or it can be corrected surgically. Undescended testes may develop malignant tumours or the higher temperature in the abdomen usually causes irreversible damage to the germinal epithelium of the seminiferous tubules, and sperm production is severely hampered leading to azoospermia (absence of sperm).

8.9.2 Congenital Inguinal Hernia

In this defect, the foetal connection between the peritoneal sac and tunica vaginalis remains permanent. The condition is also known as persistent processus vaginalis that can be surgically corrected.

8.9.3 Hydrocele

In this disorder, a clear or straw-coloured fluid accumulates within the tunica vaginalis. Such defect may be part of a general oedema, or may result from lesions of the testicle such as inflammation or tumour that usually leads to bleeding (haematocele).

8.9.4 Torsion of Testicle

It may occur during sleep and without obvious cause, and produces a twist of the cord at the inguinal ring. As a result, the testes become hard and swollen and extremely painful. Obstruction of the vein results in gross congestion and interstitial haemorrhage.

8.9.5 Eunuchoidism

The condition is related to male hypogonadism, specifically when there is a Leydig cell deficiency dating from childhood as a result of which the circulating level of androgens is depressed. Such individuals over the age of 20 are characteristically tall because their epiphyses remain open and some growth continues past the normal age of puberty. They have feminine body configuration with high-pitched voice.

8.9.6 Hirsutism

It is the disorder related with excessive growth of body hair that results from increased secretion of testosterone or its precursor androstenedione. The condition may be androgen-dependant or independent. Androgen dependant hirsutism is restricted to the chin, upper lip, chest, and other androgen-sensitive areas, while these areas as well as the forehead, abdomen, arms, and legs are involved in androgen-independent hirsutism.

8.10 REPRODUCTIVE CYCLE IN FEMALE

Mammals include diversified and highly evolved group of vertebrates (such as primates like human, apes, and monkeys, and non-primates like lion, tiger, deer, elephant, cattle, sheep, pig, etc.). Their females are well marked by the sexual cycles as compared to males. On the basis of the annual reproductive activities, two different patterns of sexual behaviour are seen in mammals and include i) *Continuous breeders*, which reproduce throughout the year (e.g. human, guinea pig, rat) and ii) *Seasonal breeders* who are interested in performing sexual activities only in specific seasons (e.g. dog, cat, sheep, cattle). Both type of breeders show cyclic pattern in their sexual behaviour. The females of mammals other than primates show a sexual cycle called the *oestrous cycle*. Oestrous is a conspicuous period of the cycle when ovulation (release of ovum from the ovary) occurs and the animal is said to be *in heat*.

It is the only period in which the female is physiologically and psychologically ready for sexual act and hence the female permits copulation only during oestrous. As compared to this, the females of primates show such period of heat after attaining puberty and is called the *menstrual cycle* marked by the periodical discharge from their genitalia. This process is called *menstruation*. Both oestrous and menstrual cycles are under the influence of the pituitary hormones, FSH and LH, and the ovarian hormones, oestrogen and progesterone. Details of the reproductive cycles are discussed below.

8.10.1 Oestrous Cycle

It is the reproductive cycle exhibited by the females of all mammals except primates. It includes the period of heat during which sexual interest is aroused. The duration of oestrous cycle varies from animal to animal and can be affected by factors like light, temperature, nutritional status and social relationships. The length of the cycle, the duration of heat and the time of ovulation in relation to the onset of heat or in induced ovulators, to the time of copulation is shown in table 8.2.

Table 8.2   Length of sexual cycle and duration of heat in some mammals.

Animal	Length of cycle	Duration of heat
Bitch	3–4 months	7–10 days
Cow	21 days	13–17 hours
Dog	2 heats/year	7–9 days
Ewe (female sheep)	16 days	30–36 hours
Guinea pig	16 days	6–11 hours
Hamster	4 days	20 hours
Mare	20–22 days	4–6 days
Rat	4–5 days	10–15 hours
Sow (female pig)	20–22 days	2–3 days
Woman	28 days	Continuous

The oestrous cycle is divided into the following four stages:

Proestrous   It is a preparatory or building up stage, the duration of which varies from animal to animal. It is associated with maturation of ovarian follicles and secretion of oestrogen from their theca interna cells. As a result, the size of the follicle-enclosing ovum increases and accumulates more follicular fluid containing oestrogen. Uterus becomes contractile and vaginal epithelium proliferates and grows thicker. A sanguinous (red) fluid oozes out through the genitalia. During this period mating is not allowed. The vaginal smear shows several nucleated epithelial cells that are detached from the proliferating vaginal epithelium (figure 8.10a).

Oestrous   As mentioned earlier, this is the period of heat that is dominated by ovulation and increased level of oestrogen released from the ruptured Graafian follicle under the influence of FSH and LH. As a consequence, mating is permitted only in this period. The other changes in the reproductive tract are related to further thickening of uterine mucosa, and the vaginal epithelium becomes cornified. The duration of heat period varies considerably in different groups of mammals. The vaginal smear is characterized by the presence of large-sized and fully keratinized epithelial cells (figure 8.10b).

Metaoestrous   This stage follows the heat period and since it occurs shortly after the release of ovum from the Graafian follicle, this stage is also called postovulatory phase. The significant feature of this phase is the formation of functional corpus luteum that is responsible for release of progesterone. As a result, there is increase in progesterone and decrease in oestrogen. The duration of this phase varies from animal to animal. Metaoestrous is also associated with diminished vascularity and contractility of uterus. The vaginal and uterine epithelial linings become thin and are infiltrated by leucocytes (figure 8.10c).

Dioestrous   It is the resting interval of relatively short period of quiescence for those mammals that shows several heat periods in a year (polyoestrous mammals). During this stage

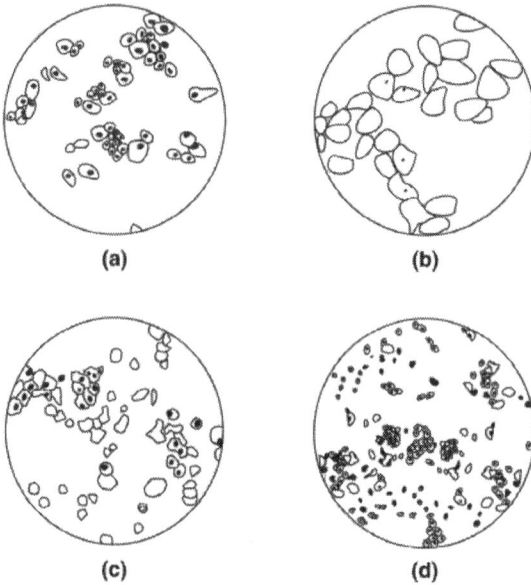

Figure 8.10   Histological features of vaginal epithelial cells observed in vaginal smear during oestrous cycle. (a) Proestrous (b) Oestrous (c) Metaoestrous (d) Dioestrous.

the animal undergoes preparatory changes for the initiation of second oestrous. The duration of this phase also varies from animal to animal. The phase is marked by regression of corpus luteum, decrease in size of ovary and uterus, diminished blood supply to uterus and the thinning of vaginal mucosa with large number of infiltrating leucocytes. Vaginal smear shows the presence of numerous leucocytes (figure 8.10d).

8.11 FEMALE REPRODUCTIVE SYSTEM

The sexual organs of women (figure 8.11) unlike that of men, shows regular cyclic changes that may be regarded as preparations for fertilization and pregnancy. In humans and other primates, the cycle is a menstrual cycle, the conspicuous feature of which is the periodic vaginal bleeding that occurs with the shedding of the uterine mucosa (menstruation). The female reproductive organs are divided into two groups—the *external genitalia* that are collectively known as vulva consist of labia majora, labia minora, the clitoris, the vestibule, the hymen, the greater vestibular glands and other *internal genitalia* that lie in the pelvic cavity and consist of the vagina, the uterus, two uterine or fallopian tubes and two ovaries.

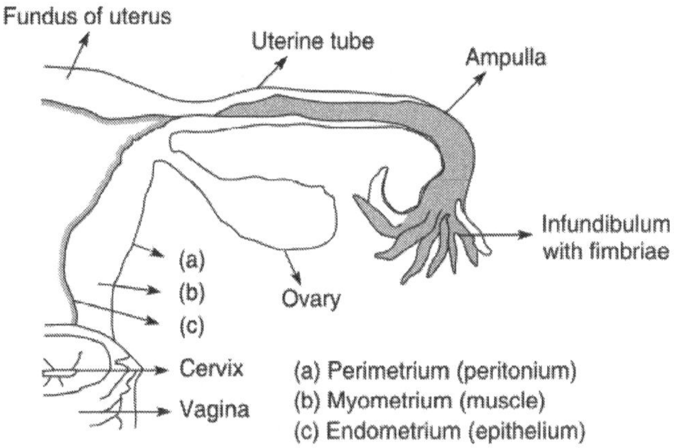

Figure 8.11 Reproductive system of a woman (only one side shown).

The *vagina* is a fibromascular tube connecting the internal and external organs of the reproductive system. It has three layers of tissues—an outer areolar, middle smooth muscular and inner stratified squamous epithelium showing transverse folds, or rugae. Autonomic nervous system and the pudendal nerves innervate the vagina playing an important role in orgasm and vaginal contractions that may help in sperm transport.

The *uterus* is a pear-shaped hollow muscular organ that lies in the pelvic cavity between the urinary bladder and the rectum. It is described in three parts viz. the fundus, the body, and the cervix. The *fundus* is the dome-shaped part of the uterus above the openings of the uterine tubes. The *body* is the main part of the uterus that is continuous with the cervix. The *cervix* protrudes through the anterior wall of the vagina and opens into it. Three layers of tissue are present in the wall of uterus, namely, an outer peritoneum (a covering of peritoneum), middle myometrium (smooth muscle fibres), and inner endometrium (a lining of mucus membrane). After puberty the uterus undergoes a regular cyclic changes, which prepares it to receive, nourish and protect a fertilized ovum. It provides the environment for the developing foetus during the gestation period.

The *uterine* or *fallopian tubes* are present one on each side of the uterus. It is divided into three parts—the *uterine part* which is present within the wall of the uterus; the *isthmus* which is the straight narrow part just lateral to the wall of the uterus; and the *ampulla* which is the widest part of the tube. The fallopian tube ends into the *infundibulum,* as a dilated trumpet-like portion opening into the peritoneal cavity, having finger-like projections called *fimbriae.* The tubes help in transport of ovum from ovary to uterus and provide the site for fertilization of the ovum.

The *ovaries* are the female gonads that are present in a shallow fossa on the lateral wall of the pelvis, and are attached to the posterior layer of the broad ligament by a band of peritoneum called mesovarium. Histologically, two distinct layers of tissues, the outer cortex and inner medulla form ovaries. The medulla is the centre of the ovary that is composed of fibrous tissue and contains blood vessels and nerves. It is surrounded by the cortex, which is composed of a network of connective tissue called stroma and is lined by germinal epithelium (figure 8.12). The cells derived from germinal epithelium undergo the process of oogenesis to form mature follicles enclosing the ovum.

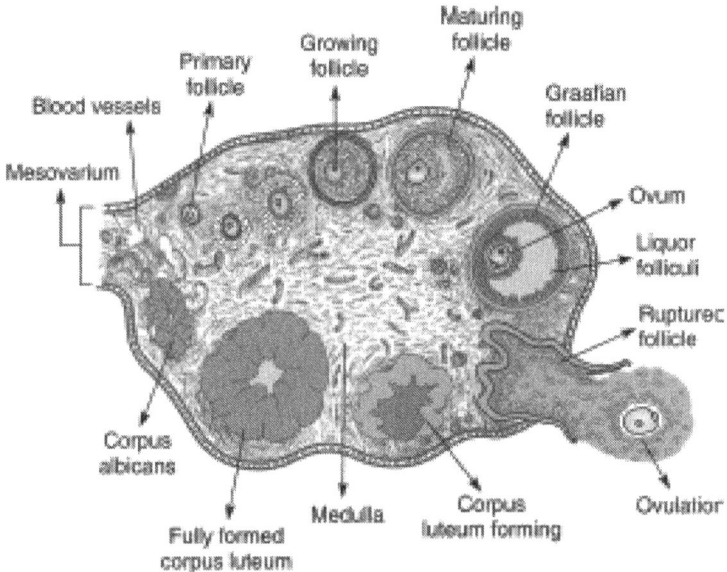

Figure 8.12   A section of ovary showing follicles at various stages of development.

A layer of epitheliod cells called granulosal cells surrounds each ovum in the ovary, and these along with the enclosed ovum are called *primary follicle*. At puberty, when the ovaries are stimulated by gonadotropic hormones, FSH and LH, secreted from anterior pituitary, a few of the primary follicles begin to enlarge. Finally the matured follicle called Graafian follicle enclosing an ovum, a cavity called *antrum* and the fluid called *liquor folliculi* is formed. A fluid-filled mature Graafian follicle measures about 10 mm in diameter and contains an ovum of about 100 to 150 μ diameter (figure 8.13). The Graafian follicle soon balloons outward on the surface of the ovary and ruptures. As a result, an ovum covered by a mass of granulosal cells is expelled into the abdominal cavity, this is the process called *ovulation*. Once the ovum is expelled into the abdominal cavity, it is picked up by the fimbriae of infundibulum and transported into the ampulla of the fallopian tube where it is ready for fertilization by the sperm.

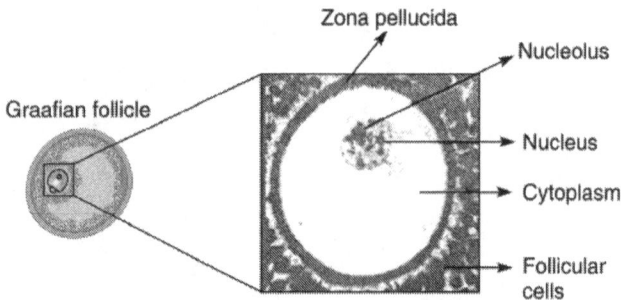

Figure 8.13   Graafian follicle with enclosed ovum (magnified).

8.11 OOGENESIS

It is the process of formation of female gamete (the ovum) from the primordial germ cells derived from the germinal epithelium in the ovaries. Figure 8.14 shows the summarized events in the process of oogenesis. In humans, no new ova are formed after birth. During foetal development, the ovaries contain over 7 million primordial follicles. However, many of them undergo atresia (involution) before birth and others are lost after birth. At the time of birth, approximately 300,000 ova are present in two ovaries. Only one of them per cycle (or about 500 in the course of a normal reproductive life of a woman) normally reaches maturity, while the rest of ova degenerate. Just before ovulation, the *primary follicles* containing *primary oocytes* complete the first meiotic division. One of the daughter cells, the *secondary oocyte*, receives most of the cytoplasm, while the other, the *first polar body*, may or may not divide, and disappears. Approximately, when the follicle ruptures, the secondary oocyte begins second meiotic division, but this division is arrested at metaphase and is completed only when a sperm penetrates the oocyte. During fertilization, the *second polar body* is cast off and the fertilized ovum is processed for the development of the foetus.

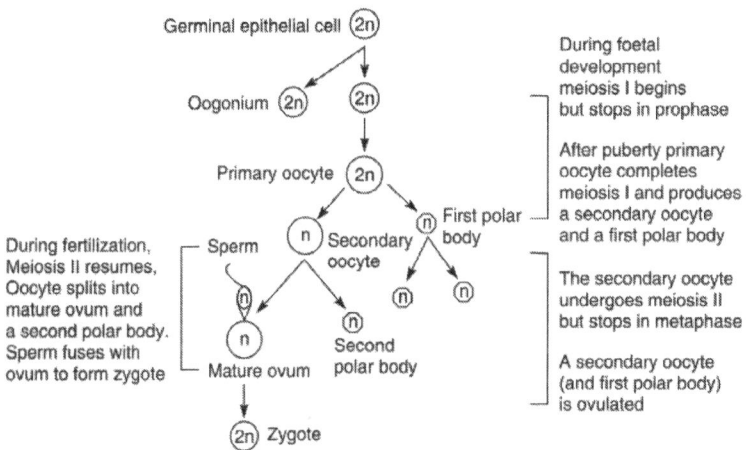

Figure 8.14   The summary of events in the process of oogenesis.

8.12 ROLE OF HORMONES IN FEMALE SEXUAL CYCLE

The sexual cycle in female is under the influence of the hormones secreted by anterior pituitary gland and ovaries. Anterior pituitary secretes two important gonadotropins— follicle stimulating hormone (FSH) and luteinizing hormone (LH). The secretion of both hormones is controlled by the hypothalamic factor or hormone called Gonadotropin releasing hormone (GnRH). The principal female sex hormones secreted by ovaries are oestrogen and progesterone. The fluctuating plasma concentrations of oestrogen and progesterone as well as FSH and LH during different phases of the monthly female sexual cycle are shown in the figure 8.17.

8.12.1 Hormones of Adenohypophysis

FSH and LH released from the anterior pituitary gland have promising role to perform in maturation of the follicle and formation of corpus luteum. At the onset of puberty, the adenohypophysis at first starts secreting FSH that initiates the beginning of the sexual life in the growing female child. Later on adenohypophysis releases LH as well, which regulates the monthly female cycle.

8.12.2 Follicle Stimulating Hormone (FSH)

It is responsible for the early maturation of the ovarian follicles. It promotes very rapid proliferation of the epithelioid cells surrounding the ovum. These cells, in turn, start to secrete oestrogen. Thus FSH performs two significant functions, first is to stimulate proliferation of ovarian follicular cells and second is to cause secretary activity by these cells. As a result, follicular cavities become well developed and there is great enlargement in the size of the follicles. Once the follicles grow to about one-half their maximum size, the adenohypophysis increases the output of luteinizing hormone along with the follicle stimulating hormone.

8.12.3 Luteinizing Hormone (LH)

This hormone along with FSH is responsible for the final maturation of the ovarian follicle. It increases still more secretion by the follicular cells, making one of the follicles so large that it ruptures, expelling its ovum into the abdominal cavity. A burst of LH secretion (*LH surge*) is responsible for this ovulation (figure 8.16). After ovulation, individual cells of the empty Graafian follicle grow in size and develop a fatty, yellowish appearance. These cells are then called *lutein cells* that collectively enlarge to form *corpus luteum*. LH further stimulates corpus luteum to continue to secrete oestrogen along with increased output of progesterone.

8.12.4 Ovarian Hormones

Oestrogen and progesterone are the two principal female sex hormones that are responsible for the sexual development and monthly sexual changes in the female. Chemically, both are steroid hormones like that of the hormones of adrenocortex and testosterone and are derived from the fatty substance cholesterol.

Oestrogen   Oestrogen is actually several different hormones that are recognized as *17β estradiol, estrone*, and *estriol.* All are C_{18} steroids with identical functions and almost but not exactly identical chemical structures (figure 8.14). Hence they are considered together as if they were a single hormone that is secreted by the granulosa cells of the ovarian follicles, the corpus luteum and placenta. The biosynthetic pathway involves formation from androgens (*Testosterones, DHT*, and *androstenedione*). Androstenedione is catalysed by the enzyme *aromatase* and converted to estrone. The same enzyme also catalyses the conversion of testosterone to 17β estradiol. Both, 17β estradiol, and estrone act as precursors for the synthesis of estriol.

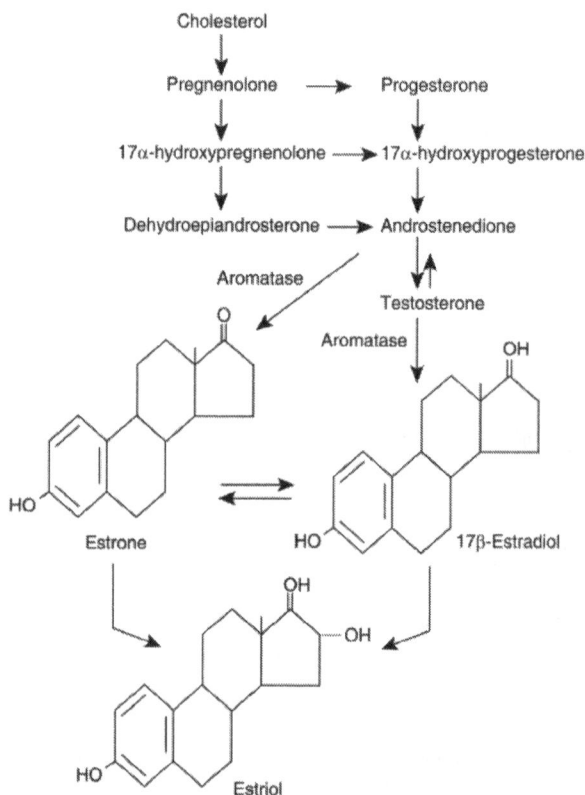

Figure 8.15 Biosynthetic pathways of oestrogens (17β estradiol, estrone, and estriol).

The ovary, as well as the testes and adrenal, has the ability to synthesize oestrogens—17β estradiol, estrone, and estriol from the androgens. During the follicular phase of the menstrual cycle (that will be discussed in later part of this chapter), ovarian secretion represents only one-third of the total oestrogen production. In contrast to estradiol, which is secreted almost entirely by ovaries, most estrone is derived from peripheral conversion of androstenedione and from estradiol metabolism. Estradiol, estrone, and estriol are bound to sex hormone binding globulin (SHBG), the same carrier protein that binds testosterone. In serum, estradiol is largely in conjugated form and is bound to sex hormone binding globulin (SHBG). In contrast, most serum estrone is present as estrone sulphate. Estradiol is the most potent of oestrogens and is present in concentrations <50 pg/ml (18.4 pmol/l) in the preovulatory period. Concentrations rise during the second half of the follicular phase and reach a peak of 150–500 pg/ml (550 to 1836 pmol/l) on the day prior to or the day of the LH surge.

Functions of Oestrogens

1. It brings about all the changes that occur during puberty in the female. The *female secondary sexual characters* that develop in girl at puberty (in addition to enlargement of breast, uterus, and vagina) are due to oestrogens. The feminine body configuration that includes narrow shoulders, broad hips, thighs that converge, arms that diverge, distribution of fat in breast and buttocks, high-pitched voice, less body hair and more scalp hair, and pubic and axillary hair, all these changes are brought into a girl who attains puberty under the influence of oestrogens. Thus oestrogens are feminizing hormones.

2. At puberty, oestrogens stimulate the growth of the ovarian follicles and increase the motility of the uterine tubes. They play an indispensable role in the cyclic changes in the endometrium, cervix, and vagina. They increase uterine blood flow and have significant effects on the smooth muscles of the uterus, making the uterus about two to three times larger than that of the child. Estrogens also increase the amount of contractile proteins in the uterine muscle.

3. *Effect on secretion of the FSH*   There is a reciprocal relationship between the oestrogens and the FSH. Estrogens decrease the secretion of FSH from the anterior pituitary gland. Women are sometimes given large doses of oestrogens for 4–6 days during their fertile period (as a part of contraception). This prevents pregnancy probably due to the inhibitory action of oestrogens over the secretion of FSH. However, administration of large doses of oestrogens for longer periods may cause atrophy of the ovaries, arrest menstruation and produce sterility.

4. *Effect on role of progesterone and corpus luteum*   For the menstrual changes and breast enlargement, both oestrogen and progesterone act synergistically. Progesterone can bring about the premenstrual changes in the endometrium only after the oestrogens have already done the proliferative changes. Similarly, the glandular development of breast, as seen during pregnancy, is because of their combined effect. Along with progesterone, oestrogens are also essential for the maintenance of the corpus luteum. However, at the time of parturition, both hormones have antagonistic effects on the wall of the uterus.

5. Oestrogens exert synergistic action with oxytocin for increasing the sensitivity and motility of uterine muscles to initiate the process of parturition.

6. *Effect on the growth of bones*   Oestrogens increase the growth rate of all bones immediately after puberty by inducing the positive calcium balance. However, they cause epiphysial closure and hence stop growth. As a result, the female grows very fast for the first few years after puberty and then ceases growing entirely.

7. *Effect on nervous system*   The sexual desire of the female in the animals other than the primates is the oestrous or the period of heat and increase in libido in humans are the impacts of oestrogens. They apparently exert this action by a direct effect on certain neurons in the hypothalamus.

8. Oestrogens help in retention of salt and water, increase of blood volume and of the water content of the muscle. However, there is increased level of aldosterone in the luteal phase of the sexual cycle in female that may contribute to this retention.

9. The secretions of the sebaceous gland become more fluid under the influence of oestrogens and hence it counteracts the effect of testosterone and inhibits the formation of comedones (blackheads) and acne.

10. Oestrogens have significant role in lowering the plasma cholesterol level and they rapidly produce vasodilation by increasing the local production of NO (see chapter 1). These actions inhibit the artherogenesis and contribute to the low incidence of myocardial infarction and other vascular diseases in postmenopausal women.

11. Oestrogens are anabolic hormones and hence they bring about nitrogen retention (protein synthesis) in the body. However, this effect is considerably less as compared to the anabolic effect of testosterone.

Progesterone   Progesterone is a C_{21} steroid hormone secreted by the corpus luteum, placenta (during pregnancy), and also in small amounts by follicles and testes. It acts as an intermediate in biosynthetic pathways for steroids in all tissues that secrete steroid hormones. In ovaries, it is synthesized from cholesterol and pregnenolone (figure 8.16). The small amounts of progesterone in male and non-menstruating females are derived from extraglandular conversion of adrenal pregnenolone and pregnenolone sulphate to progesterone, and by secretion of progesterone by the adrenals. In serum, about 18 per cent of progesterone is bound to cortisol-binding globulin and 79 per cent to albumin, whereas the remainder is free (unbound).

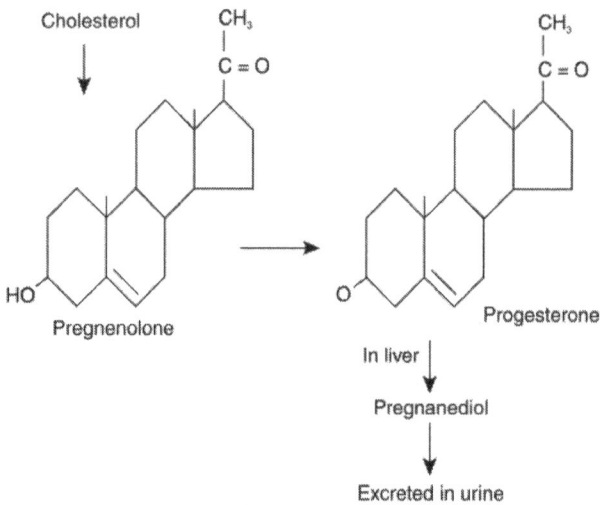

Figure 8.16   Biosynthetic and metabolic pathway of progesterone.

Progesterone production by the corpus luteum can be indirectly assessed by measurement of the basal body temperature. During the luteal phase there is about a 0.5°C rise in the body temperature that lasts about 10 to 12 days and parallels the increase in progesterone concentration. With the radioimmunoassay methods, the serum level of progesterone can be measured and that is <1 ng/ml (3.2 nmol/l) during preovulatory phase. At the time of the LH surge, serum progesterone levels begin to rise and about 4 to 6 days later reach a peak of about 10 to 20 ng/ml (31.8 to 63.6 nmol/l). After remaining more or less stable for about one week, its concentrations drop rapidly to about 1 ng/ml a short time before the onset of menstruation. In the major pathway of its metabolism, progesterone is converted in the liver to pregnanediol, which is conjugated to glucuronic acid and excreted in the urine.

Functions of progesterone It has little to do with the development of female secondary sexual characters; instead it is principally involved in the preparation of the uterus for implantation of the zygote and preparation of the breast for secretion of milk. Its biological actions are as follows:

1. *Effect on uterus and placenta formation* Progesterone is responsible for the pro-gestational changes in the uterus. Under its influence, endometrium becomes considerably thick and vascular so that uterus prepares itself for the reception and embedding of the fertilized ovum and also makes it possible to convert the uterine mucosa into placental membrane. If corpus lutea are removed after implantation of the zygote, placenta could not form and hence embryo dies which leads to abortion.

2. It exerts antagonistic effect to oxytocin on the uterine muscle. Once the fertilized ovum is implanted and placenta is developed, the corpus lutea and placenta secrete copious amounts of progesterone, which neutralizes the effect of oxytocin and prevents the muscular contractility of the pregnant uterus thereby reducing the chances of abortion. Thus progesterone is essential for maintenance of pregnancy.

3. Progesterone has an anti-oestrogenic effect on the myometrium and thus decreases the excitability of the uterus and its sensitivity to oxytocin.

4. In the breast, progesterone promotes the development of lobules and alveoli. It stimulates differentiation of oestrogen-prepared ductal tissue and is responsible for secretary function of the breast during lactation.

5. Large doses of progesterone prevent LH secretion and also exert inhibitory action on oestrous or menstrual cycle, maturation of follicle and ovulation.

6. Progesterone brings about the relaxation of pelvic ligaments and enlarges the birth canal in order to accommodate the developing foetus.

7. It does not have significant anabolic action but has mild water and salt retention effect. Administration of large doses of progesterone in animals produce natriuresis, probably by blocking the action of aldosterone on the kidney.

8. Progesterone stimulates the process of respiration. The alveolar PCO_2 in women during the luteal phase of the menstrual cycle is lower than that in men.

9. It has thermogenic action. As mentioned earlier, it is probably responsible for the rise in basal body temperature by 0.5°C at the time of ovulation and in postovulatory phase.

8.12.5 Relaxin

In addition to the above-mentioned hormones, there is a hormone called relaxin, which is secreted by corpus luteum, uterus, placenta and mammary glands in women and by the prostate gland in men. Chemically, it is a water-soluble polypeptide hormone. During pregnancy, this non-steroidal hormone relaxes the pubic symphysis and other pelvic joints and softens and dilates the uterine cervix. Its plasma concentration reaches the peak during the last week of pregnancy. Thus, relaxin facilitates parturition. Its function in non-pregnant women is unknown. In men, relaxin is found in semen, where it may maintain the motility of sperm and help the sperm to penetrate the ovum.

8.13 MENSTRUAL CYCLE

This is the sexual cycle that characteristically occurs in primates only and where copulation may be performed at any time after puberty. As mentioned earlier, in oestrous cycle there is a period of heat during which the female permits mating, whereas in menstruating primates including women, there is no such heightened sexual desire. The puberty in girl (usually commences at the age of 12–14 years) is characterized by the first menstruation, the *menarche,* and its most conspicuous feature is the periodic vaginal bleeding consisting of blood, mucus and certain other substances that occurs with the shedding of the uterine mucosa (menstruation). The length of the cycle varies considerably. On an average it takes 28 days (24–30 days) in physiologically normal women. Menstruation is a periodical phenomenon that continues from puberty to menopause. With the advancing age of the woman, the ovaries become unresponsive to gonadotropins, and they lose their function thereby causing the sexual cycle to disappear; this is the age (45–55 years) at which the reproductive function of the woman comes to an end and this stage of her life is called menopause. Menstrual cycle is absent, before puberty, during pregnancy and after menopause. For the convenience of description, menstrual cycle is divided into four phases that are discussed in the following sections.

8.13.1 Proliferative Phase

This phase is also known as follicular phase or repair phase since it involves the maturation of the ovarian follicle and repair of the endometrium after menstruation. The period starts from the sixth day of menstruation and lasts for 10 to 12 days to repair the damage caused to the uterus during the previous mensus. The phase is accompanied by proliferation of the endometrial lining with increasing blood supply and accumulation of ground substances that result in increase in thickness of the uterus. In the ovary, the ovarian follicle grows to form a matured Graafian follicle. The proliferative phase is under the influence of follicle stimulating hormone secreted by the anterior pituitary and the oestrogen by follicular cells. All the related changes in the wall of the uterus, in the ovary, and the hormonal levels in plasma during the menstrual cycle are shown in the figure 8.17.

8.13.2 Ovulatory Phase

This is the significant phase of the menstrual cycle that occurs at about the 14th day of the intermenstrual cycle or midway between two menstrual periods. There is no remarkable change in the endometrial thickness during this phase, but it is principally marked by the rupture of the Graafian follicle. As a result, the ovum is expelled into the abdominal cavity and there is rise in the basal body temperature by 0.5°C that remains high until the onset of the next menstrual period. There is a surge in luteinizing hormone (LH) secretion from the anterior pituitary that triggers ovulation. The event of ovulation normally occurs about 9 hours after the peak of the LH surge at midcycle (see figure 8.17).

8.13.3 Secretory or Premenstrual or Luteal Phase

It starts from the 15th day of the menstrual cycle and lasts for 11 to 13 days, during which the empty Graafian follicle is

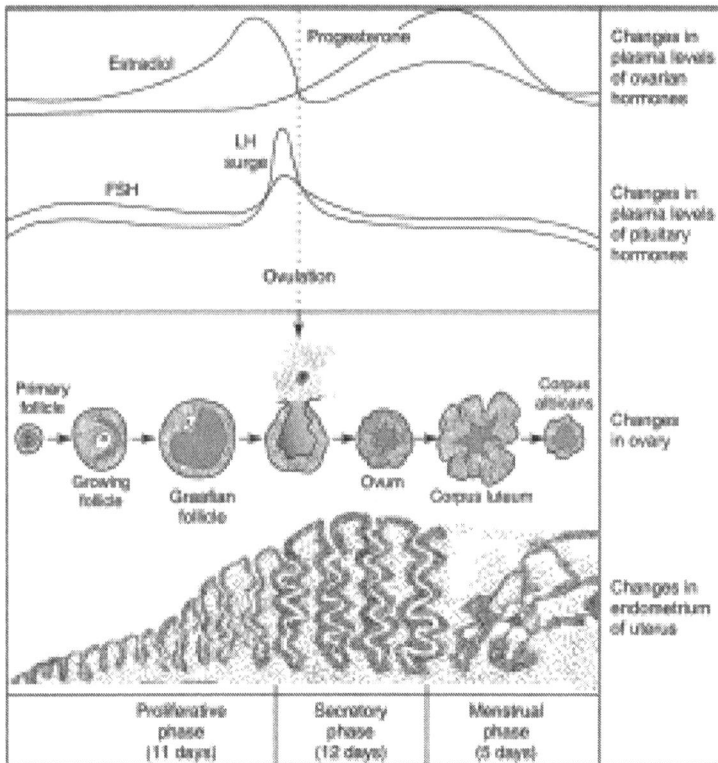

Figure 8.17 Menstrual cycle showing changes in plasma concentrations of the pituitary and ovarian hormones along with periodical changes in ovary and endometrium of the uterus.

developed into fully formed corpus luteum. The corpus luteum secretes copious amount of progesterone and small amount of oestrogen. Pro-gesterone brings about the pro-gestational changes in the oestrogen-prepared endometrium so that the thickness of the endometrial lining is further increased with more vascularization. The uterine glands become elongated, much coiled and secrete more mucus. Large quantities of fatty substances and glycogen are deposited in the deeper cells of endometrium. As a consequence, uterus appropriately prepares itself for the implantation of the fertilized ovum. If pregnancy occurs, corpus luteum persists, otherwise it degenerates before menstruation.

8.13.4 Destructive Phase or Menses

This phase is marked by the vaginal bleeding and shedding of endometrial mucosa that lasts for 4–5 days. If fertilization takes place the fertilized ovum is implanted in the uterus and in due course of development the zygote (blastocyst) itself starts secreting the hormone called *chorionic gonadotropin*. This hormone keeps the corpus luteum intact helping it to continue the secretion of progesterone for the first 3 to 4 months of pregnancy. However, if there is no fertilization and implantation of the ovum, no such hormone is produced. As a consequence, corpus luteum disappears and the plasma concentrations of oestrogen and progesterone drastically decrease. In the absence of these hormones, the endometrial blood vessels constrict and the tissue dies and sloughs into the uterine cavity, this process is known as menstruation. About 50 ml of blood, necrotic endometrial tissue, mucus, numerous leucocytes and unfertilized ovum, is gradually expelled from the genital tract.

The menstrual phase is again followed by the proliferative phase under the influence of renewed oestrogen secretion from the ovaries and thus the sexual cycle continues.

8.14 MENOPAUSE

The reproductive period of a woman lasts for about 35 years. Menopause is the term given to the period, when the ovaries become unresponsive to the FSH and LH secreted from anterior pituitary, which may occur between the age of 45–55. During this period the ovulations and sexual cycles become irregular and finally disappear. The reduced responsiveness of the ovaries may be associated with the decrease in the number of primordial follicles. Menopause may also be associated with other phenomena like: i) gradual atrophy in uterus and vagina with advancing age, ii) increased plasma concentrations of FSH and LH in the absence of negative feedback from oestrogen and progesterone, iii) vasodilation with hot flushes (sensations of warmth spreading from the trunk to face), night sweating, palpitations and disturbances in normal sleep pattern, iv) shrinkage of breast, and v) other associated psychic problems.

Although the ability of testes to produce sperm and testosterone tends to decrease slowly as the age advances, men do not undergo 'male menopause' or andropause.

8.15 ABNORMALITIES IN MENSTRUAL CYCLE

Amenorrhea is the condition in female where there are no menstrual periods. In *primary amenorrhea*, menstrual bleeding has never occurred. Some women with such condition have small breasts with other signs that include failure to mature sexually. Some women may stop menstruating for the rest of their reproductive life after parturition (the process of birth). Such cessation cycles in a woman with previously normal periods is called *secondary amenorrhea*. There are various causes of amenorrhea that include pregnancy, emotional stimuli, environmental changes, hypothalamic diseases, pituitary defects, and abnormalities in ovarian functions.

Hypomenorrhea and *menorrhagia* are the terms assigned to scanty and abnormally profuse flow respectively, during regular menstrual cycles. *Dysmenorrhea* is the painful menstruation characterized by several menstrual cramps that are common in young women, and quite often disappears after the first pregnancy. Dysmenorrhea is due to accumulation of prostaglandins in the uterus, where symptomatic relief can be obtained by treatment with inhibitors of prostaglandin synthesis.

Premenstrual syndrome (PMS) is the condition which occurs in some women due to retention of salt and water in the body, and is characterized by irritability, bloating, odema, depression, headache and constipation during the last 7–10 days of their menstrual cycles. Treatment with antidepressant drugs can produce symptomatic relief.

Thinning of vulval epithelium, malignant vulvar tumours, viral infection to vagina, cervical ectopy (red area on ectocervix), cervical intra-epithelial neoplasia, acute and non-specific chronic endometritis, endometrial hyperplasia, benign and malignant tumours of endometrium and myometrium, damaged fallopian tube (salpingitis), and non-neoplastic cysts and adenocarcinomas in ovaries, are diseased conditions of female reproductive tract, which severely affect the respective function.

In addition to the above-mentioned abnormalities, there are genetic defects that arise due to mutation in a single gene and lead to reproductive defects in women. These include (a) Kallmann's syndrome, which causes hypogonadotropic hypogonadism; (b) GnRH resistance, FSH resistance, and LH resistance, which are due to defects in the GnRH, FSH, and LH receptors, respectively; and (c) Aromatase deficiency, which prevents formation of oestrogens.

8.16 PREGNANCY AND ENDOCRINE CHANGES

Pregnancy or gestation is the period from the moment of fertilization till parturition (act of giving birth). In the physiologically normal woman pregnancy lasts for 280 days, which corresponds to the period for 10 menstrual cycles. Pregnancy stops ovulation and menstrual cycles during this period because placenta secretes large amounts of oestrogens and progesterone that in turn inhibit the secretion of FSH and LH from the anterior pituitary due to feedback mechanism. Since FSH and LH are the only hormones responsible for the maturation and ovulation of Graafian follicle, their absence in the blood during pregnancy stops ovulation.

8.16.1 Fertilization

Once the ovum is expelled from the ovary, it remains viable for 8 to 24 hours. In humans, fertilization of the ovum by the sperm usually occurs in the ampulla of the fallopian tube within 12 hours of coitus. Millions of sperms are deposited in the vagina during intercourse. Eventually, 50–100 sperms are attracted to the ovum by substances produced by the ovum (chemoattraction). Many of them contact the zona pellucida (the membranous structure surrounding the ovum) through the sperm receptors. This is followed by the acrosomal reaction, i.e., the breakdown of acrosome of the sperm and release of several enzymes including trypsin-like protease *acrosin*, which facilitates the penetration of sperm through the zona pellucida (figure 8.18). After a single sperm has reached the membrane of ovum, fusion to ovum membrane is mediated by *fertilin*, a protein present on the surface of the sperm head. The fusion of sperm head (a male pronucleus) with the nucleus of the ovum (female pronucleus) provides the signal for initiation of the development. The fusion also triggers the physicochemical reaction in the membrane of the fertilized ovum to become impermeable to the entry of additional sperm, and thus prevents polyspermy (fertilization of the ovum by more than one sperm). The developing embryo, is now called *blastocyst*, passes along the fallopian tube into the uterus. This journey takes about 3 days, during which the blastocyst reaches the 8- or 16-cell stage. In the course of development, the blastocyst

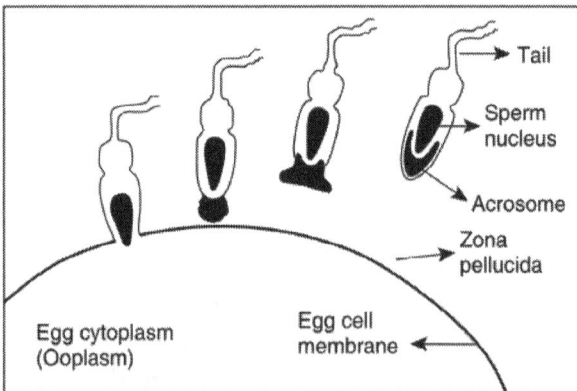

Figure 8.18 Diagrammatic representation of the sequential events in fertilization of the ovum.

undergoes the process of cell differentiation and develops an *inner cell mass* that will eventually form the embryo and outer cell membrane called *trophoblast* that eventually forms the foetal membranes.

8.16.2 Implantation of Developing Embryo

When the developing blastocyst, having a cellular mass of about 16–32 cells, comes in contact with the progesterone-prepared endometrium (that eventually attains thickness of about 10 mm), the trophoblast further differentiates into an outer layer

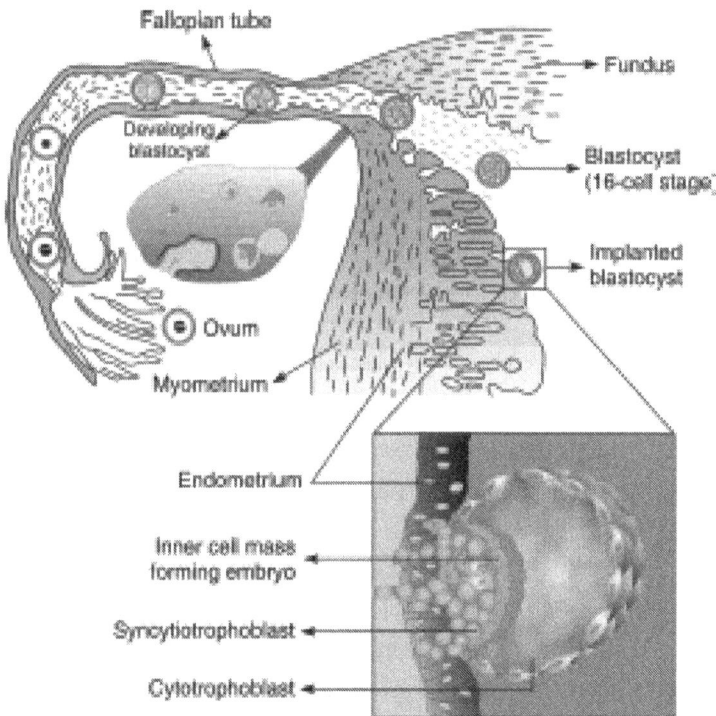

Figure 8.19 Process of implantation of the blastocyst in uterus. (Implanted blastocyst magnified).

of *syncytiotrophoblast*, a multinucleate mass with no distinct cell boundaries surrounding the blastocyst, and an inner layer of *cytotrophoblast* made up of individual cells. The cells of the outer layer secrete large quantities of proteolytic enzymes that digest the endometrium, and then trophoblasts derive their nutrition by phagocyting the endometrial cells that have stored large quantities of proteins, lipids and glycogen. At the same time, the blastocyst burrows into the endometrium constituting the process called *implantation* (figure 8.19). Usually, the blastocyst is implanted on the dorsal wall of the uterus. The trophoblasts grow and divide very rapidly and soon they, along with the adjacent cells, begin to form placenta and foetal membranes while the embryo develops inside the blastocyst from the inner cell mass.

8.17 PLACENTA

It is a functional connection between the developing embryo and the wall of the uterus (mother) that permits nutritional, respiratory and excretory interchange of the material by simple diffusion. Placenta also prevents the entry of invading bacteria and other large molecules into foetal circulation. Figure 8.20 shows a pregnant uterus with developing foetus, foetal membranes and placenta. The foetal membranes form the amniotic cavity that is filled by amniotic fluid. The foetus floats freely in the amniotic fluid, which also protects the foetus from mechanical shocks. The placenta occupies about one-sixth surface of the uterus. The foetus is connected with the placenta by the umbilical cord that transports the blood enriched with nutrients from the placenta to the baby. At the same time it helps in transporting the excretory products from the foetal circulation to the placenta and subsequently to the maternal blood.

8.17.1 Placental Hormones

The placenta produces sufficient amount of hormones to take over the function of corpus luteum after the sixth week of pregnancy. For the maintenance of pregnancy and other pregnancy-related physiological changes, the placenta secretes the following hormones:

i. *Human chorionic gonadotropin (hCG)* It is a hormone secreted by the syncytiotrophoblast. Chemically, it is a glycoprotein that contains galactose and hexosamine. Like pituitary glycoprotein hormones, it is made up of α and β subunits. The alpha subunit of hCG is identical to α subunit of LH, FSH, and TSH. The molecular

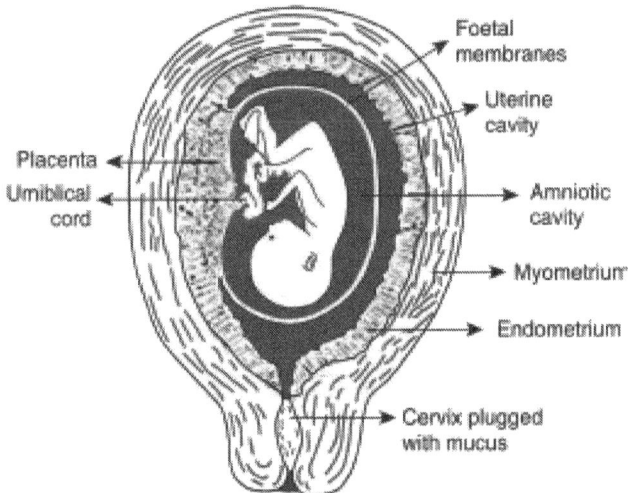

Figure 8.20 The pregnant uterus showing foetus, foetal membranes and placenta.

weight of hCG-α is 18,000 and that of hCG-β is 28,000. It can be measured by radioimmunoassay and detected in the blood as early as 6 days after conception. Its presence in the urine in early pregnancy is the basis of various laboratory tests for pregnancy.

During the early days of pregnancy, hCG stimulates the corpus luteum to enlarge and secrete oestrogens, progesterone and relaxin. Thus, hCG has primarily a luteinizing and luteotropic activity and little FSH activity.

ii. *Human placental lactogen (hPL)* or *chorionic growth hormone prolactin (CGP)*   This hormone is also secreted by syncytiotrophoblast with lactogenic and small amount of growth-stimulating activity and hence called hPL or CGP respectively, but now it is generally known as *human chorionic somatomammotropin (hCS)*. Chemically, it is also a protein hormone having very similar structure to that of human growth hormone and apparently, hCS functions as a 'maternal growth hormone of pregnancy' and it brings about the retention of nitrogen, potassium and calcium. At the same time, hCS helps in diverting the glucose to the foetus and also stimulates the growth of milk producing structures of breasts and actually causes milk-production after the breasts have been appropriately prepared by oestrogens and progesterone. These are very much similar effects that the hormone prolactin, secreted by the anterior pituitary gland, has on the breasts after the birth of the baby.

iii. *Relaxin* or *uterine-relaxing factor (URF)*   The sources of this hormone are placenta, uterus and ovaries. It helps in relaxation of the ligament of the symphysis pubis and inhibits myometrial contraction.

iv. *Progesterone and oestrogens*   These hormones are secreted by syncytiotrophoblast.

v. Other hormones secreted by the human placenta are *β-endorphin, β-MSH, dynorphin A, leptin, prolactin,* and *prorenin.*

All the above-mentioned hormones secreted by human placenta bring about the following physiological changes during pregnancy:

1. Stimulation of corpus luteum for growth and persistence in early days of pregnancy

2. Inhibition of ovulation

3. Development of placenta

4. Enlargement of birth canal and relaxation of pelvic ligaments that allow the pelvic opening to stretch as the baby is born

5. Enlargement of breasts, pigmentation of areola and nipple

6. Increase in cardiac output and vital capacity of lungs

7. Enlargement of thyroid and parathyroid glands

8. Metabolic changes including glycosuria, retention of nitrogen, lipolysis, and accumulation of water in amniotic fluid, placenta, uterus, blood, and breast.

9. Nausea and vomiting associated with early days of pregnancy

10. Enlargement of foetal adrenal glands and increased output of steroid hormones

11. The pregnant mother gains weight (approximately 10 kg.) during pregnancy due to the developing foetus, enlarged birth canal, placenta, foetal membranes and

breast, increased fat deposition in various parts of the body and increased amount of body fluids.

8.18 ABNORMALITIES RELATED TO PREGNANCY

Some of the important aspects of pregnancy-related abnormalities are discussed in the following sections.

8.18.1 Hydatidiform Mole

It is an abnormal conceptus (the product of conception) in which an embryo is absent and the placental mass is often unduly large and distends the uterine cavity; the placental villi appear as clusters of grapes. Such condition always leads to abortion. This disorder has a relationship with *choriocarcinoma* (a highly malignant neoplasm of the trophoblast). The woman with such abnormality secretes large amounts of the placental human chorionic gonadotropin (hCG), hence it is desirable for the patients who have had a mole to be periodically assessed by estimation of their urinary hCG output.

8.18.2 Ectopic Pregnancy

It is an extra-uterine pregnancy in which a fertilized ovum implants and begins to develop before it reaches its natural site in the uterus. About 97 percent of ectopic pregnancies occur in fallopian tubes, most commonly in ampullary region. In such disorder, the placenta cannot be fully formed, the invading trophoblast weakens and fallopian tube may rupture leading to intraperitonial bleeding. Clinically the condition is associated with acute abdominal pain that demands surgical treatment.

8.19 PARTURITION

It is the process of giving birth to the baby, which normally occurs in humans at an average of 40 weeks from the beginning of the last menstrual cycle. There are chances of survival of the babies that are born as early as 28 weeks or as late as 46 weeks. Approximately 90 per cent of all babies are born within 10 days before or after the 40-week interval.

There is still considerable uncertainty about the mechanism responsible for the onset of the birth process. The probable factors that cause onset of parturition include progressive mechanical as well as hormonal changes, which are as follows:

1. When the baby becomes large, it exerts pressure on the uterus and initiates uterine muscular contraction.

2. Movements of the baby like striking the feet and hands against the uterine wall also cause contraction in uterine muscles.

3. The cervix remains firm in the nonpregnant state and throughout pregnancy, but at the time of delivery it becomes soft and dilates, while the uterus contracts and the foetus is expelled out of the birth canal.

4. The plasma concentration of progesterone (which prevents uterine contractions) secreted by the placenta drops near term. As soon as progesterone level is decreased during the end of gestation, labour starts.

5. One of the important factors responsible for the onset of parturition is the increase in circulating oestrogens shortly before birth that makes the uterus more excitable and causes production of more prostaglandins, which in turn cause uterine contractions.

6. The hormone oxytocin secreted by hypothalamic–posterior pituitary system shortly before the end of pregnancy also plays an indispensable role in initiation of vigorous uterine contractions.

8.19.1 Mechanism of Parturition

The act of giving birth to baby starts as labour pains (i.e. onset of progressively strong uterine contractions), and ends in the actual expulsion of the baby (parturition). The circulating levels of progesterone and oestrogens play significant role in onset of labour. During pregnancy, under the influence of progesterone, the uterus remains in a quiescent state. But in the last 3 months of the pregnancy, the uterine contractions become progressively strong and reach extreme intensity a few hours before birth because of the cumulative effect of decreasing plasma concentration of progesterone and increasing circulating oestrogens, oxytocin, prostaglandins and relaxin. And finally the baby is expelled from the birth canal.

The period of labour is variable, on an average it lasts for 12 to 18 hours. During the *first stage* or *the stage of dilation*, the head of the baby pushes against the cervix and acts as a wedge to dilate and open the cervical and vaginal canals. In most of the cases the head of the baby stretches the cervix while in some cases the legs come out first and this makes the act of delivery more difficult. During the first stage, the amnion is ruptured and amniotic fluid is released through the birth canal. During the *second stage* or *stage of descent*, the head of the baby descends first and then the remaining body slips within a few seconds through the dilated vaginal canal. This is due to the cervical stretch that excites the fundic contraction, which in turn pushes the baby down and stretches the cervix some more. The cycle repeats over and over again resulting in expulsion of the baby from the birth canal (figure 8.21). Finally, in the *third stage* or *placental stage*, the foetal membranes are expelled, thus brigaing the process of parturition to an end.

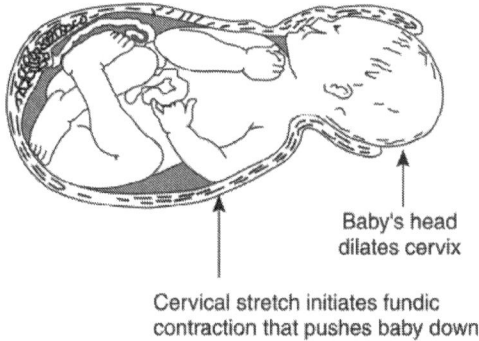

Baby's head
dilates cervix

Cervical stretch initiates fundic
contraction that pushes baby down

Figure 8.21   The mechanism of parturition.

In addition to the role played by oestrogens, progesterone and relaxin in the act of parturition, oxytocin significantly participates in the process of giving birth to a baby. Once labour is stared and cervix is dilated, this dilation in turn sets up signals in afferent nerves that stimulate hypothalamic–posterior pituitary system to increase oxytocin secretion. As a result, the plasma concentration of oxytocin increases and more oxytocin is made available to act on the uterus. This hormone directly acts on the uterine muscle and makes them to contract. In addition, it also stimulates synthesis of prostaglandins in the smooth muscle cells of the endometrium that promotes the oxytocin-induced contraction. Thus, there is the sort of positive feedback mechanism that helps in the process of delivery and expulsion of the baby from the birth canal (figure 8.22).

Parturition is followed by the after-birth process called *involution* in which the uterus rapidly decreases its size by gradual autolysis and resumes the original position in the pelvic cavity and becomes approximately its original size. The process requires 6 to 8 weeks. The hypertrophied myometrial and endometrial smooth muscle cells undergo self-destruction so that it soon resumes its original size and shape.

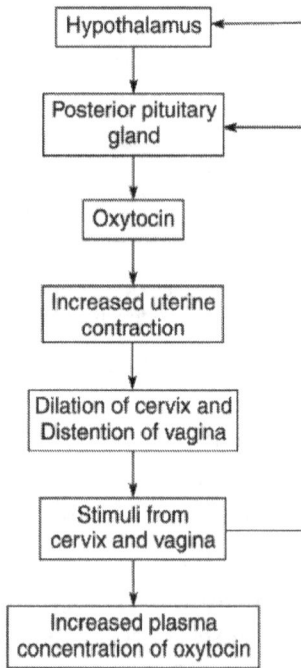

Figure 8.22　Participation of oxytocin in parturition through positive feedback mechanism.

8.20 LACTATION

It is the synthesis and ejection of milk from the breasts for nursing the baby. Lactation is a complex process that requires a neat balance between hormonal, structural, nervous, and emotional factors.

8.20.1 Development of Breasts (Mammary glands)

The female reproductive system consists of the breasts as accessory reproductive structures. Mammary glands are rudimentary in child and males. Occasionally, breasts may be developed in male producing the condition called *gynecomastia*, which can be found in the patients with oestrogen-secreting tumour, and in conditions like eunuchoidism, hyperthyroidism, and liver cirrhosis. The common cause of the gynecomastia is either increase in plasma concentration of oestrogens or decrease in circulating androgens.

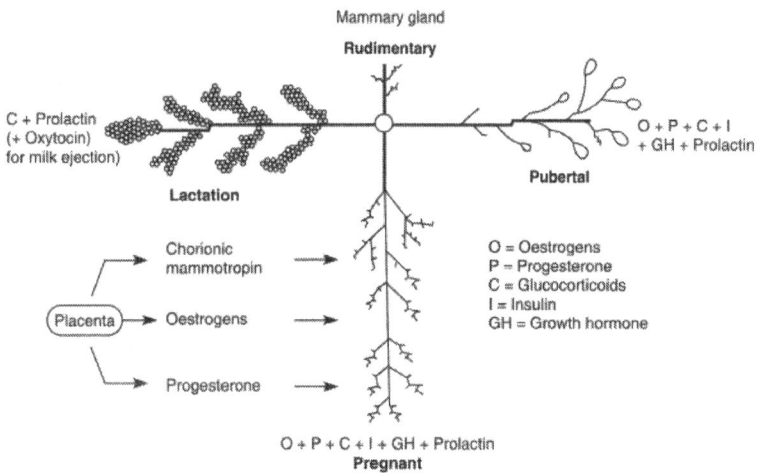

Figure 8.23 Hormonal effect on development of mammary gland and lactation at various stages viz., pubertal (right), pregnant (below) and lactation (left).

In female, the breasts begin to develop at puberty. Structurally, breast consists of numerous lobules and clusters of alveoli, the secretory cells of the breast, from which minute lactiferous ducts pass towards the nipple. The development of breast is primarily under the influence of ovarian hormones. Oestrogens induce proliferation of the lactiferous ducts and progesterone promotes the development of the lobules in the breasts. In addition to ovarian hormones, insulin, glucocorticoids, growth hormone, prolactin and oxytocin also have influencing role in the development of the breast lactation and ejection of milk (figure 8.23). All of these hormones are essential for proliferation and growth of breasts at puberty. During pregnancy, full lobuloalveolar development of breasts and some milk secretion takes place due to cumulative effect of the hormones—oestrogens, progesterone, glucocorticoids, insulin, growth hormone and prolactin.

8.20.2 Secretion and Ejection of Milk

The breasts enlarge and glandular elements begin to develop in response to higher plasma concentration of oestrogens, progesterone, prolactin and hCS. Some milk is secreted into lactiferous ducts. In woman, at the time of parturition, sudden expulsion of placenta from the uterus removes the source of oestrogens and progesterone. As a result, the circulating levels of oestrogens and progesterone abruptly decline. During gestation, placental oestrogens and progesterone have inhibited the production of prolactin by anterior pituitary gland, but after parturition these hormones are not available, hence anterior pituitary starts to secrete large quantity of prolactin. In the presence of prolactin, all the secretory cells of the breasts produce copious amount of milk.

Ejection of milk from the breast is the synchronized action of hormonal, nervous and emotional factors. When a baby sucks on the nipples, milk flows through nipples as result of milk ejection reflex. The suckling by the baby causes sensory nerve signals to pass into spinal cord, then to brain stem and finally to hypothalamus that produces oxytocin. This hormone circulates through the blood to breast, where it stimulates *myoepithelial cells* surrounding the alveoli to contract, with consequent ejection of milk through ducts leading to the nipples (figure 8.24).

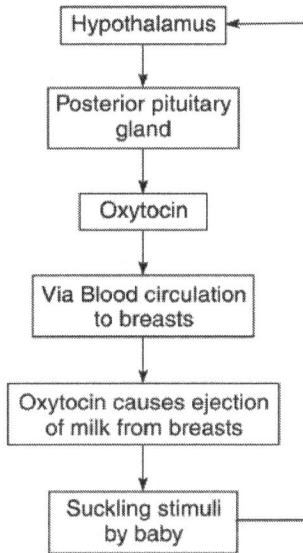

Figure 8.24 Milk ejection reflex.

Milk ejection mechanism in lactating women can be adversely affected by emotional factors like worry, sadness, and fear of disturbing their figure. Also, disturbances caused by other children in family may inhibit milk flow; as a result breasts fail to empty that in turn prevent the anterior pituitary to secrete prolactin, and breasts cease to secrete milk.

8.20.3 Composition of Milk

Milk is considered to be a complete food as it contains all the substances needed for energy and growth of the infant. It contains proteins in the form of casein and lactalbumin, lactose as easily digested sugar, and large quantity of fat. Milk also contains small quantity of vitamins and large quantities of calcium phosphate. The relative compositions of human and cow's milk are shown in Table 8.3. Newborns that are not nourished with human milk, should be preferably fed with cow's milk. In such condition, it is desirable to fortify cow's milk with extra quantity of pure glucose (dextrose), as it is more easily digestible than cane sugar (sucrose).

Table 8.3 The composition of human and cow's milk

Constituents	Human milk	Cow's milk
Water (g/dl)	88.5	87.0
Lactose (g/dl)	6.8	4.8
Fat (g/dl)	3.3	3.5
Protein (g/dl)	1.7	3.3
Casein:lactalbumin ratio	1 : 2	3 : 1
Calcium (mg/dl)	33	125

8.21 EFFECT OF LACTATION ON MENSTRUAL CYCLE

During pregnancy, the breasts are prepared to synthesize and eject milk under the impact of the synchronized action of several hormones. The prime objective of this preparation is to nurse the baby after parturition. But in some women who do not nurse their infants, menstrual period may recur 6 weeks after parturition. However, lactating mothers who nurse their babies regularly have amenorrhea as early as 8 weeks or as late as 30 weeks after delivery, depending on the plasma concentration of prolactin. As long as the mother continues to lactate with copious amount of prolactin secreted by anterior pituitary gland, it is evident that prolactin prevents the production of FSH and LH from the anterior pituitary gland. As a consequence, ovaries remain inactive, ovulation is inhibited, and the rate of synthesis of oestrogens and progesterone is suppressed. Hence, the percentage of women becoming pregnant again during the suckling period is considerably negligible.

Chiari-Frommel syndrome is an interesting although rare disorder in women who do not nurse their infants after delivery and show persistent lactation (galactorrhea) and amenorrhea. The condition may be associated with atrophy of genital organs and due to continuous production of prolactin without the secretion of gonadotropins, which are necessary for maturation of new ovarian follicle and its subsequent ovulation.

SYNOPSIS

❑ Reproduction is the perpetuation of the species that occurs either by *asexual* or *sexual* method.

❑ Sexual reproduction involves meeting of two parents or the union of cells from each parent followed by fertilization, pregnancy, parturition, and lactation. All these events are under the hormonal influence.

❑ There are several abnormalities in sexual development of human beings, which could be genetic or hormonal abnormalities.

❑ In chromosomal abnormalities, an established defect in gametogenesis is meiotic nondisjunction that may lead to *Turner's syndrome* or gonadial dysgenesis, *Klinefelter's syndrome* or seminiferous tubule dysgenesis, and 21-trisomy or *Down's syndrome* or Mongolism.

❑ A *pseudohermaphrodite* is an individual with the genetic constitution and gonads of one sex and genitalia of the other. The condition can be either female pseudohermaphroditism or male pseudohermaphroditism depending on the availability of androgen.

❑ Mutation in the androgen receptor gene forms the basis for androgen resistance that leads to *testicular feminizing syndrome* or complete androgen receptor syndrome.

❑ *Puberty* or *adolescence* is the final maturation of gonads in both the sexes under the influence of gonadotropins secreted by anterior pituitary.

❑ *Precocious puberty* is an early development of secondary sexual characteristics without spermatogenesis or oogenesis in immature males and females when they are abnormally exposed to androgen and oestrogen respectively.

❑ Spermatogenesis in testes is initiated by gonadotropic hormones (i) *follicle-stimulating hormone (FSH)* and (ii) *luteinizing hormone (LH) or interstitial cells stimulating hormone (ICSH)* that stimulates Leydig cells in testes to secrete the male sex hormone, *testosterone* responsible for development of secondary sexual characteristics.

❑ Chemically, testosterone is a C_{19} steroid synthesized from *cholesterol* in the Leydig cells and also from *androstenedione* secreted by the adrenal cortex.

❑ Testicular abnormalities include undescended testes (*cryptorchidism*), congenital inguinal hernia, hydrocele, torsion of testicle, *eunuchoidism* (male hypogonadism), and androgen-dependant or independent hirsutism.

❑ *Continuous breeders* reproduce throughout the year while *seasonal breeders* perform sexual activities only in specific seasons.

❑ The females of mammals other than primates show sexual cycle called *oestrous cycle* that is marked by a *period of heat* during which the female shows desire for mating. The cycle includes proestrous, oestrous, metaoestrus and dioestrous phases.

❍ *Menstrual cycle* is the sexual cycle of primates only, where the copulation may be performed at any time once they attain puberty. It consists of four phases viz., proliferative, ovulatory, secretory or premenstrual or luteal and destructive or menses. All the phases are under the influence of the hormones secreted by anterior pituitary gland and ovary.

❍ Abnormalities in menstrual cycle may lead to amenorrhea (absence of menstrual periods), hypomenorrhea, menorrhagia, dysmenorrhea and premenstrual syndrome (PMS).

❍ Fertilization is followed by implantation of fertilized ovum (blastocyst) having inner cell mass (actual embryo forming cells) and trophoblast, which further differentiates into an outer layer of *syncytiotrophoblast*, and an inner layer of *cytotrophoblast*.

❍ The functional connection between the foetus and the wall of the uterus (mother) is the *placenta* that permits nutritional, respiratory and excretory interchange of the material.

❍ Placenta secretes human chorionic gonadotropin (hCG), human placental lactogen (hPL) or chorionic growth hormone prolactin (CGP) or human chorionic somatomammotropin (hCS), relaxin or uterine-relaxing factor (URF), progesterone and oestrogens.

❍ Abnormalities related to pregnancy are *hydatidiform mole, choriocarcinoma* and *ectopic pregnancy*.

❍ *Parturition* is the process of giving birth to the baby, which is completed in three stages—*first stage* or *the stage of dilation* (of birth canal), *second stage* or *stage of descent* (of baby through vaginal canal) and *third stage* or *placental stage* (the expulsion of foetal membranes). Parturition is based on synchronization of mechanical and hormonal factors.

❍ *Lactation* is the synthesis and ejection of milk from the breasts for nursing the baby, that requires a neat balance between hormonal, structural, nervous, and emotional factors.

○ *Milk ejection reflex* involves suckling by the infant and release of oxytocin from hypothalamic-posterior pituitary system that results into ejection of milk from the breasts.

○ *Human milk* contains casein, lactalbumin, lactose, fat, small quantity of vitamins and large quantities of calcium phosphate.

○ In lactating mothers, *prolactin* prevents the production of FSH and LH from anterior pituitary gland, ovaries remain inactive, and ovulation is inhibited.

○ *Chiari-Frommel syndrome* is a disorder in women who do not nurse their infants after delivery, showing persistent lactation (galactorrhea) and amenorrhea.

REVIEW QUESTIONS

1. What are the differences between asexual and sexual reproduction?

2. Define nondisjunction of chromosomes. Explain the consequences of meiotic nondisjunction.

3. Describe hermaphroditism. What are the different types of hermaphroditism?

4. What is puberty? Explain pubertal changes in both male and female.

5. Trace the development of sperm and its maturation and describe structure of sperm.

6. Explain the role of follicle stimulating hormone and luteinizing hormone in male.

7. Mention the biosynthetic pathway for male sex hormone. In what ways does testosterone affect bodily function?

8. Explain various testicular abnormalities.

9. What are different sexual cycles in female? Explain oestrous cycle with its various phases.

10. Define continuous breeders and seasonal breeders. What are the physiological changes in animals during heat period?

11. Explain the parts of birth canal with the structure of ovary.

12. Trace the formation of Graafian follicle and explain the process of ovulation.

13. What are the functions of follicle stimulating hormone and luteinizing hormone in female?

14. Explain the biosynthetic and metabolic pathway of oestrogens with its functions.

15. How does corpus luteum form? What are the effects of progesterone on the body?

16. Describe menstrual cycle in detail with the changes in plasma concentration of ovarian and pituitary hormones.

17. Describe endometrial cycle and menstruation.

18. Explain various abnormalities related to menstrual cycle.

19. In what way does sperm fertilize the ovum and explain the process of implantation of fertilized ovum.

20. How does placenta develop? Describe the various roles of placenta including its endocrine functions.

21. Define gestation. What are the functions of various hormones during pregnancy?

22. What are the changes that occur in the mother during pregnancy?

23. Explain various abnormalities associated with pregnancy.

24. What is parturition? Explain the various stages in the act of delivering a baby.

25. How do the mechanical and hormonal factors interact during parturition?

26. Explain the role of various hormones in the preparation of breasts and add a note on milk ejection reflex.

27. What are the effects of lactation on menstrual cycle?

28. Write short note on:

    i.   Budding

    ii.   Androgen resistance

    iii.   Precocious puberty

    iv.   Sex chromatin

    v.   Eunuchoidism

    vi.   Vaginal smear

    vii.   Duration of heat

    viii. LH surge

    ix.   Menopause

    x.   Kallmann's syndrome

    xi.   Aromatase deficiency

    xii.   Fertilin

    xiii. Inhibin

    xiv.   Syncytiotrophoblast

    xv.   Role of oxytocin in parturition

    xvi.   Gynecomastia

    xvii. Involution

    xviii. Composition of milk.

NINE

FEEDBACK CONTROL OF
HORMONE PRODUCTION

9.1 PLASMA CONCENTRATION OF HORMONES

The physiological effects of hormones depend largely on their concentration in blood and extracellular fluid. Almost inevitably, disease results when hormone concentrations are either too high or too low, and precise control over circulating concentrations of hormones is therefore crucial.

The concentration of hormone as seen by target cells is determined by three factors

1. *Rate of production* Synthesis and secretion of hormones are the most highly regulated aspects of endocrine control. Positive and negative feedback circuits mediate such control.

2. *Rate of delivery* An example of this effect is blood flow to a target organ or group of target cells–high blood flow delivers more hormone than low blood flow.

3. *Rate of degradation and elimination* Hormones, like all biomolecules, have characteristic rates of decay, and are metabolized and excreted from the body through several routes. Shutting off secretion of a hormone that has a very short half-life causes circulating hormone concentration to plummet, but if a hormone's biological half-life is long, effective concentrations persist for some time after secretion ceases.

Hormones need not only to be switched on at the right time, they also need to be switched off once they have produced the desired effect. A common mechanism for this is a *feedback loop*, in which the output from a hormone's action feeds back into the system as input, switching the hormone off. Feedback circuits are at the root of most control mechanisms in physiology, and are particularly prominent in the endocrine system. Instances of positive feedback certainly occur, but negative feedback is much more common.

The operation of biological and physiological systems involves *input* and *output*. In the self-regulated system, the output exerts control over the input. This is called feedback control. When we inverse this relationship, an increase in the output leads to a decrease in the input, the regulation is by *negative feedback*. When the relationship is direct, meaning an increase in output leads to a further increase in input, then it is called *positive feedback*.

9.2 NEGATIVE FEEDBACK LOOPS

They are used extensively to regulate secretion of hormones in the hypothalamic–pituitary axis. Table 9.1 shows the relationship among hypothalamic, pituitary, target glands, and feedback. Negative feedback promotes homeostasis and equilibrium within the internal environment.

Table 9.1 Relationship among Hypothalamus, Pituitary, Target Glands, and Feedback.

Hypothalamic Regulatory Hormone	Pituitary Hormone	Target Gland	Feedback Hormone
TRH	TSH	Thyroid gland	T_4, T_3
LH-RH	LH	Gonad	E_2, T
LH-RH	FSH	Gonad	Inhibin, E_2, T
GH-RH, SMS	GH	Multi-organs	IGF-1
PIF	Prolactin	Breast	(Unknown)
CRH, ADH	ACTH	Adrenal	Cortisol

ACTH = Adrenocorticotropic hormone; ADH = Antidiuretic hormone; CRH = Corticotropin-releasing hormone; E_2 = Estradiol; FSH = Follicle-stimulating hormone; GH = Growth hormone; GH-RH = Growth hormone-releasing hormone; IGF = Insulin-like growth factor; LH = Luteinizing hormone; LH-RH = Luteinizing hormone-releasing hormone; PIF = Prolactin release-inhibitory factor; SMS = Somatostatin; T = Testosterone; T_4 = Thyroxine; TRH = Thyrotropin-releasing hormone; TSH = thyroid-stimulating hormone.

Homeostasis has survival value because it means an animal can adapt to a changing environment.

Some of the important examples of *negative feedback loops* are discussed in the following sections.

9.2.1 Feedback Control of Thyroid Hormone Secretion

The thyroid hormones thyroxine and triiodothyronine ("T4 and T3") are synthesized and secreted by thyroid glands and affect metabolism throughout the body (as mentioned earlier in 3.1.3). The basic mechanisms for control in this system are:

- Neurons in the hypothalamus secrete thyroid-releasing hormone or factor (TRF), which stimulates cells in the anterior pituitary to secrete thyroid-stimulating hormone (TSH).

- TSH binds to receptors on epithelial cells in the thyroid gland, stimulating synthesis and secretion of thyroid hormones, which affect probably all cells in the body.

- When blood concentrations of thyroid hormones increase above a certain threshold, TRF-secreting neurons in the hypothalamus are inhibited and stop secreting TRF. This is an example of "negative feedback".

- Inhibition of TRF secretion leads to shut-off of TSH secretion, which leads to shut-off of thyroid hormone secretion. As thyroid hormone levels decay below the

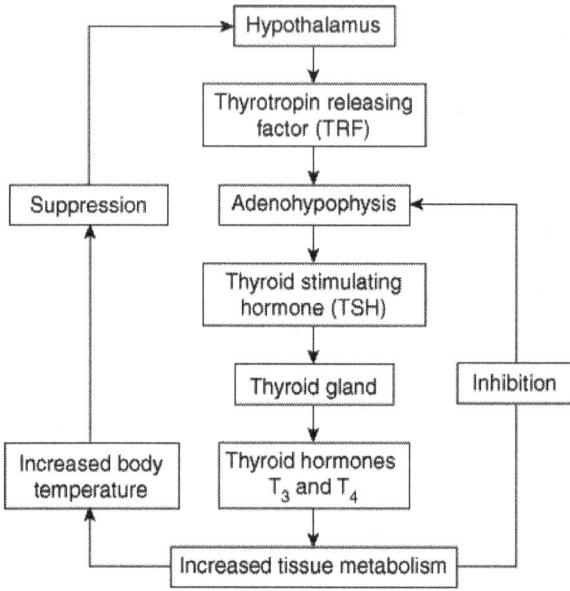

Figure 9.1   Feedback control of thyroid secretion.

threshold, negative feedback is relieved; TRF secretion starts again, leading to TSH secretion (figure 9.1).

9.2.2 Feedback Loop for ADH Secretion

This is the good example for the regulation of water levels in the blood by a hormone. The pituitary releases ADH when a part of the brain called the hypothalamus detects too little water in the blood. ADH triggers the kidney to absorb less water from the blood so less water reaches the bladder to be excreted, and the water level in the blood rises. When the hypothalamus detects that blood water level is normal, the pituitary is triggered to stop releasing ADH (figure 9.2). The hormonal mechanism of ADH is discussed in more detail in unit 2.2.17.

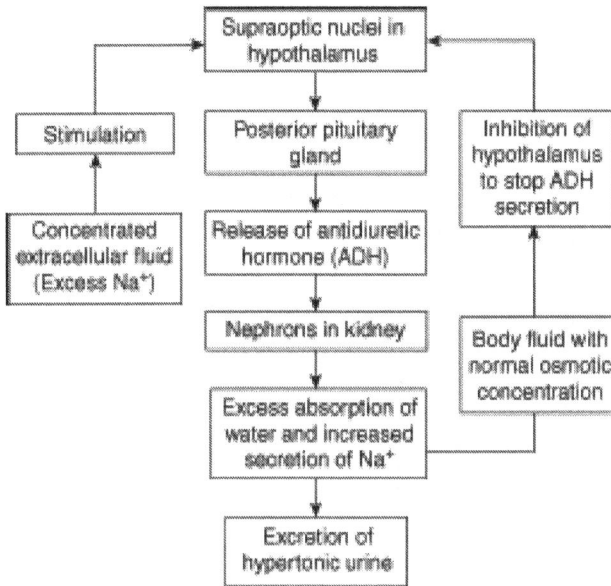

Figure 9.2 Antidiuretic hormone feedback control system.

9.2.3 Feedback Mechanism for Insulin Secretion

Another type of feedback is seen in endocrine systems that regulate concentrations of blood components such as glucose. Drink a glass of milk or eat a candy bar and the following (simplified) series of events will occur:

- Glucose from the ingested lactose or sucrose is absorbed in the intestine and the level of glucose in blood rises.

- Elevation of blood glucose concentration stimulates endocrine cells in the pancreas to release insulin.

- Insulin has the major effect of facilitating entry of glucose into many cells of the body, as a result of which blood glucose levels fall.

- When the level of blood glucose falls sufficiently, the stimulus for insulin release disappears and insulin is no longer secreted.

9.2.4 Feedback Control for the Secretion of Parathyroid Hormone

Decreased plasma calcium ion (Ca^{++}) levels trigger an increase in parathyroid hormone from the parathyroid gland, which acts on a bone to release calcium. Heightened plasma calcium inhibits further release of parathyroid hormone. Optimal plasma calcium levels are maintained by the negative feedback mechanism.

Negative feedback ensures equilibrium and stability of hormone levels. All physiological homeostatic mechanisms, including numerous endocrine glands and their hormones, operate using negative feedback. Conversely, positive feedback regulation tends to create disequilibrium, leading to abnormally high hormone conditions and disease. However, occasionally normal physiological events occur using the positive feedback mechanism. Hormones can sometimes use positive feedback between the hypothalamic regulatory mechanisms. Some examples of positive feedback are the sexual act leading to orgasm, ovulation, parturition and milk ejection from breasts due to suckling by infant.

9.3 REQUIREMENT OF NEGATIVE FEEDBACK CONTROL

Because mammals are warm-blooded, the enzymes that are part of their make-up as a warm-blooded animal require a certain temperature to operate optimally. Also, the water concentration of a cell and its chemical concentration must remain at a certain level to allow normal cellular processes to occur. Thus, the feedback mechanism in warm-blooded animals is essential for the body to work in optimal conditions.

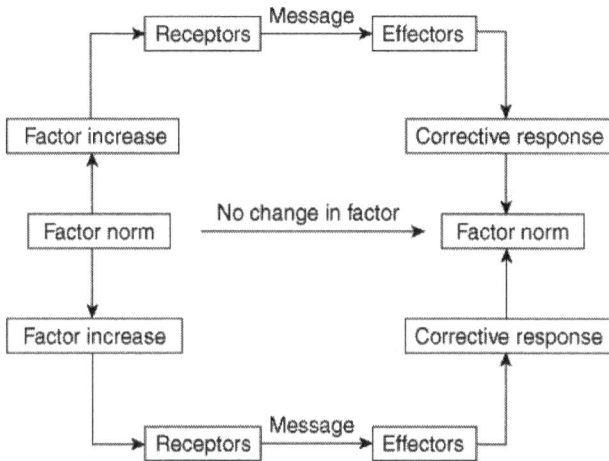

Figure 9.3   The principle of negative feedback control.

The regulation of hormone secretion in the body is achieved at different complexity levels. Simple hormonal regulation involves just one endocrine gland. The secretion of hormone from the gland is controlled directly through negative feedback, by the plasma concentration of physiological variable parameter that is regulated by hormone. The endocrine cell contains a receptor to detect the blood level of the parameter, while the complex hormonal regulation involves the anterior pituitary and the target cell. In this mechanism, one endocrine gland is controlled by the hormones of another.

Figure 9.3 illustrates the principle of negative feedback control for physiological homeostasis, translating in layman's terms to the physical equilibrium. It is essentially a corrective mechanism; consider the following scenario in a person.

- The level of glucose in the bloodstream drops.

- The person requires glucose in the cells to meet the demand for ATP.

- The body detects this with a particular receptor designed for this function.

- These receptors release hormones, specifically glucagon from alpha cells in islets of Langerhans in pancreas, which are chemical messages that initiate the start of the feedback mechanism.

- The hormones travel to their target tissue and initiate a corrective response.

- In this case, the corrective response is the secretion of more glucose into the bloodstream.

9.4 ADVANTAGES OF HOMEOSTASIS

Homeostasis has survival value because it means an animal can adapt to a changing environment. The body will attempt to maintain a norm, the desired level of a factor to achieve homeostasis. However, it can only work within tolerable limits, where extreme conditions can disable the negative feedback mechanism. In these instances, death can result, unless medical treatment is executed to bring about the natural occurrence of these feedback mechanisms.

9.5 POSITIVE FEEDBACK MECHANISM

Here the relationship between input and output of the system is direct; hence an increase in output leads to a further increase in input. As a consequence, the system changes more and more rapidly towards some extreme state whereas in negative feedback mechanism, system tends to become normal (figure 9.4). Positive feedback loops are very rare as compared to negative feedback.

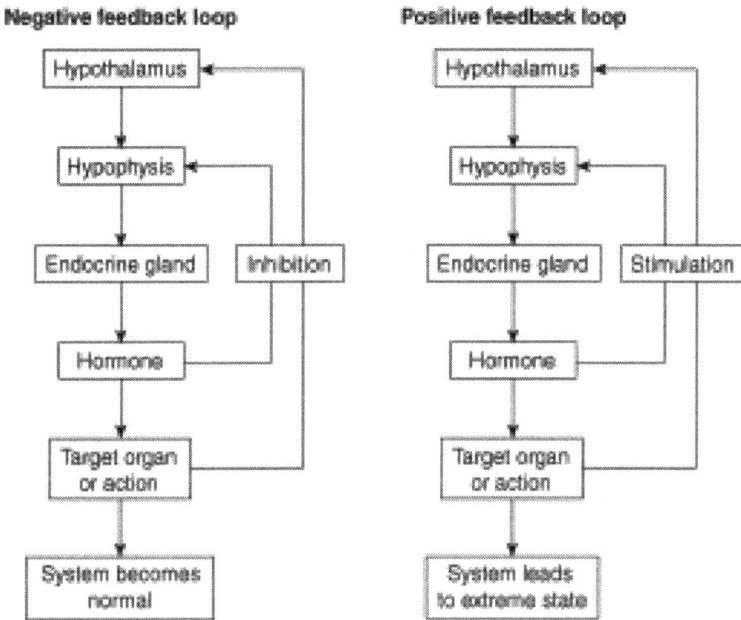

Negative feedback loop

Hypothalamus
Hypophysis
Endocrine gland — Inhibition
Hormone
Target organ or action
System becomes normal

Positive feedback loop

Hypothalamus
Hypophysis
Endocrine gland — Stimulation
Hormone
Target organ or action
System leads to extreme state

Figure 9.4 Comparison of negative and positive feedback loops.

Positive feedback mechanisms are executed by the relationship between FSH and ovulation, the parturition and release of oxytocin, the courtship between opposite partners that leads to orgasm, and the suckling stimuli and milk ejection. All these examples lead to extreme states (ovulation, parturition, orgasm, and milk ejection) that are briefly described as follows:

i. During preovulatory phase of the menstrual cycle, there is positive relationship between FSH and the secretion of oestrogens from follicular cells of the growing follicle that leads to ovulation (in presence of LH surge) at approximately 14th day of the normal sexual cycle.

ii. At the time of parturition, the uterine muscles progressively increase their contractility as the plasma concentration of oxytocin increases. In addition to mechanical changes during the act of delivering the baby, oxytocin secreted by posterior pituitary brings about vigorous contraction of pregnant uterus that leads to parturition.

iii. The courtship behaviour involves synchronizing events that are started with responses to small signals in suitable partners. It is followed by progressive mutual reinforcement and continued positive feedback between the mating partners leading to increase in desire for sexual act and finally culminates into orgasm.

iv. The suckling stimuli provided by the infant stimulates the lactating mother. In addition to several hormones, there is positive feedback mechanism that operates between oxytocin and milk ejection from the breasts. The progressive increase in oxytocin secreted by posterior pituitary gland in response to suckling stimuli leads to copious flow of milk from the breasts.

SYNOPSIS

○ Circulating concentrations of hormones is crucial for producing the physiological effects, and plasma levels of hormones largely depend upon the rate of production, the rate of their delivery to target organ and rate of degradation and elimination of the hormones.

○ The endocrine system is marked by feedback loops either of negative or positive types forming the root of most control mechanisms in physiology.

○ Negative feedback involves an increase in the output leading to a decrease in the input that promotes homeostasis and equilibrium within the internal environment. Homeostasis has survival value because it helps an animal to adapt to a changing environment.

○ All physiological homeostatic mechanisms, including numerous endocrine glands and their hormones are based on negative feedback.

○ Negative feedback loops are exemplified by the relationship of thyroid hormones with tissue metabolism, antidiuretic hormone secreted by posterior pituitary gland with the concentration of sodium and water in blood, insulin with blood glucose level, and parathyroid hormone with plasma calcium level.

○ In positive feedback mechanism the relationship between input and output of the system is direct; hence an increase in output leads to a further increase in input. As a consequence, the system changes more and more rapidly towards some extreme state.

○ Positive feedback loops are executed by the relationship between FSH and ovulation, the parturition and release of oxytocin, the sexual act of desired partners leading to orgasm, and the suckling stimuli and milk ejection.

REVIEW QUESTIONS

1. Explain the term feedback loop. Describe the types of feedback mechanism with their significance.

2. What are the points of difference between negative and positive feedback loops?

3. Describe the relationship between thyroid hormones with hypothalamus and anterior pituitary gland.

4. Explain the positive feedback mechanism with the help of various examples.

5. Write short notes on

    i.   Factors determining the plasma concentration of hormones

    ii. Homeostasis

    iii. Principle of negative feedback control

TEN

HORMONES AS PHARMACEUTICALS

10.1 HORMONES IN CONTRACEPTION

There are various methods used by humans to prevent conception, which ultimately help in controlling the population growth. Implantation of intrauterine devices (IUDs) like copper T, and Loop D form the effective mode of contraception, which sometimes can cause intrauterine infections.

Use of contraceptive pills made up of synthetic steroid hormones forms the commonest and more effective method in human females to avoid pregnancy. The pills are administered for 21 days, and then withdrawn for 5–7 days to permit menstrual flow, and started again. For contraception, an orally active oestrogen such as *ethinyl estradiol* is often combined with a synthetic progestin (that mimics the action of progesterone) such as *norethindrone*. Women treated with such contraceptive pills do not ovulate because the secretion of FSH and LH is suppressed. A synthetic oestrogen is a potent contraceptive and, unlike naturally occurring oestrogen, is relatively active when given orally, because it is resistant to hepatic metabolism. Estradiol stimulates the growth of endometrium and the breast and can lead to cancer of the uterus and, probably, the breast. As per the need of time, active research has been made in this direction to develop such derivatives of oestrogen that has bone and cardiovascular effects and not the growth stimulating effect on uterus and breast. In this connection, two compounds–*Tamoxifen* and *Raloxifene,* show promising results, since they do not stimulate the growth of uterus and breast, but both have bone-preserving and cardiovascular effects of estradiol.

In the Indian market, two improved contraceptives are available and commonly used by females to have a child by choice and not by chance. These include (i) contraceptive pills with the trade name *Saheli* developed by Central Drug Research Institute (CDRI) Lucknow that is taken once a week producing no side effects, and (ii) Contraceptive injection having trade name Depo-provera containing sterile *medroxy-progesterone acetate* developed by Pharmacia MV/SA, Belgium. It is administered once in 3 months and it does not produce any side effects.

In some part of the world, women are showing interest in using implants made up primarily of progestin such as *levonorgestrel.* These are inserted under skin and can reduce the chances of pregnancy for up to 5 years. They often produce amenorrhea, but otherwise they appear to be effective and well tolerated.

10.2 OESTROGEN AND CANCER

About 35% of carcinomas of the breast in women of childbearing age are oestrogen-dependent; their continued growth depends upon the presence of oestrogen in the circulation. Major risk factors in the development of breast cancer are estrogenic hormones in addition to ionizing radiation and other chemical mutagens. Too much estrogen; estrogen too soon, as during a young girl's development; inappropriate timing of estrogen exposure; and the wrong kinds of estrogen are factors in this breast cancer epidemic. The tumors are not cured by decreasing oestrogen secretion, but symptoms are dramatically relieved, and the tumor regresses for months or years before recurring. Women with oestrogen-dependent tumors often have a remission when their ovaries are removed.

Why would anyone produce synthetic hormones? On the surface a number of the products appear to answer a panoply of needs, and despite known negative effects on human and animal health, they have been approved by the FDA for a number of medical "ills," and sanctioned by the Agriculture Department as animal growth agents.

One of the most widely used synthetic hormones is *diethylstilbestrol*, also called, DES. DES was prescribed to women to prevent miscarriage, to treat post menopausal "complications," to "cure" headaches, dizziness, nervousness, depression, frigidity, insomnia, muscle and joint pains, vaginitis (including gonorrhea), and infertility; and it was taken to prevent conception. In developed countries, DES was administered to animals (like cattle, swine, poultry birds, etc.) to increase their weight, and subsequently, the weight of the person who consumed the meat of such animals that had been dosed with DES.

DES was and is easy and cheap to manufacture, affording large profits to those companies that sell it. As a consequence, before it was finally banned from much of the meat supply, DES became administered to nearly the entire US population.

The failure to prevent the use these chemicals, and now the remedy for the problems resulting from synthetic estrogenic compounds in the environment are not because of lack of knowledge. The history of the development of synthetic estrogens such as DES with its enormous commercial value has been known for decades. Understanding these historical developments is basic to understanding the development of today's epidemic of hormonally related cancers, and may help illuminate the enormous power, pressure, and profits behind this industry.

Hormone replacement therapy (HRT) has had the long-standing benefit of preventing bone loss in post-menopausal patients. With rising concerns of increased cancer risk with oestrogens, new selective oestrogen receptor modulators (SERMs) have been shown to have similar efficacy while not stimulating receptors in the breast or uterus, reducing risk of cancer.

10.3 SEX HORMONES AND THE IMMUNE SYSTEM

The concept of immunology has advanced from a mere response to infectious agents to include a variety of complex situations and interactions with most, if not all, of the body's systems. For the purpose of understanding these interactions they may be divided into "classical" immune regulation and non-immune regulation by immunocytes. We will deal mainly with non-immune regulation and its interaction with hormone replacement therapy (HRT).

The elements of non-immune regulation include stimuli, effector cells, signals and target cells. Each of the components of non-immune regulation is sensitive to the reproductive milieu and therefore to gonadal function. Chief among these are the circulating monocytes that arise in the bone marrow and are the precursors to the tissue macrophages. In many organs estrogen regulates the number of tissue macrophages. Further, during menopause the number of circulating monocytes increases.

Thus gonadal steroids represent primary signals for non-immune regulation. Acting via steroid receptors, they can regulate monocyte number, cytokine production by monocytes and differentiation of monocytes into macrophages in the tissues. Monocytes contain estrogen receptors, but macrophages contain both aromatase and estrogen receptors. They respond to estradiol by secretion of cytokines, which can act in an autocrine or paracrine manner to regulate cell number and cell function.

It has long been suspected that there is a strong interaction between sex hormones and the immune system. Several observations support this concept such as the dimorphic nature of the two genders; alterations in the immune response after gonadectomy or sex steroid hormone treatment, modification of the immune response during pregnancy, and the identification of steroid hormone receptors in cells of the immune system support this concept. Numerous *in vitro* and *in vivo* experiments have demonstrated that sex hormones affect and modify the actions of cells of the immune system. In addition, considerable evidence has accumulated suggesting that the interaction between estrogen and cells of the immune system can have non-immune regulatory effects. Thus, the role of estrogen in the prevention of bone loss is mediated by mechanisms involving the inhibition of pro-inflammatory cytokines by bone marrow cells. Moreover, disorders frequently affecting women after menopause, such as cardiovascular disease, osteoporosis, or neuro-degenerative disorders, can be ascribed to the loss of sex hormone-dependent regulation of physiological functions, as well as to a modification of the non-immune functions of resident immune cells.

10.4 DHEA

DHEA (or Dehydroepiandrosterone) is a natural sterone produced by the adrenal gland and is in fact a metabolite of cholesterol. DHEA is the most common sterone in human blood. But amounts decline rapidly with age. Amounts are highest during the early twenties and begin to decline at around age 25, by the time we reach 70 years of age, DHEA production is only a small fraction of what it was 50 years earlier. Research has shown a correlation between low DHEA levels and a declining immune system, and DHEA is being used in the fight against HIV, cancer and senile dementia. It is also been clinically shown that DHEA helps brain neurons establish contact. It is known that Alzheimer patients have low DHEA levels when compared to their healthy counterparts. It is also known that a small amount of sulphate DHEA and micronized DHEA is converted into testosterone. But DHEA's most over-looked but vital role may be its relationship with cortisol. Research is indicating that DHEA is the counter-balance to cortisol, e.g. when DHEA is low, cortisol levels are high and vice-versa.

Cortisol is one of the few hormones that increase with age, it is known to induce stress and when allowed to circulate at high levels for long periods may affect many bodily functions, including insulin resistance and damage to the endocrine system via damage to the hypothalamus. Maintaining healthy levels of DHEA for aging and stressed individuals may be most important, because of DHEA's ability to help lower cortisol levels. The average pulsate production of DHEA from healthy adrenal glands is approximately 25 mg per day, less for some women and more for some men. It is likely that many of the dosages of DHEA being used today (50–100 mg daily) are too great for long-term continuous use, and although there is no known down-regulation (a situation whereby the adrenal glands would slow or stop their own manufacture of DHEA in response to the continuous high levels caused by long-term DHEA supplementation), it is always advisable to stop DHEA use for periods of time on a regular basis, to prevent this possibility (or better still have the DHEA levels monitored). An answer may be to use only low doses of DHEA on a regular basis.

Liquid micronized DHEA is probably only required for normal aging individuals at a dosage of 1 to 10 drops daily (average of 3–5 daily). Each drop being 4 mg sublingual micronized DHEA, and because of the improved absorption properties of sublingual administration each drop could be equivalent to about double or quadruple the same amount of capsule DHEA. Liquid sublingual micronized DHEA may be more ideal in the administration of small doses of DHEA over long periods.

7-keto DHEA can be viewed as the final metabolite of DHEA breakdown and may be more "responsible" for DHEA's immune enhancing effects. 7-keto DHEA is also known for not converting to androgens (and hence not to testosterone or estrogen also) and may be more suitable for those wishing to avoid that effect. This may be particularly beneficial for women, (especially those in the menopause) who often complain of "testosterone side-effects" such as facial hair and acne when using "other" forms of DHEA. 7-keto DHEA as a final metabolite is believed to be two and a half times more potent than other forms of DHEA. Doses of 7-keto DHEA are 25 mg to 50 mg daily with occasional breaks. DHEA should not be used by persons who suffer from, have suffered from or may be suffering prostate or testicular diseases including cancer.

10.5 PROGESTERONE

Women often use progesterone cream in the menopause especially to "counter-weight" hormone replacement therapy (HRT- estrogen therapy). Unfortunately many physicians who prescribe HRT forget about progesterone. When they do prescribe progesterone it is often the alien progesterone derived from horse urine and discovered in the 1920's. It is now easy and relatively cheap to synthesize natural human progesterone, and this progesterone cream is just that- natural progesterone.

Clinical studies show that whilst oestrogen levels drop dramatically during menopause they level out afterward and the body continues to produce oestrogen (albeit in much smaller quantities). But whilst progesterone levels decline less rapidly they continue to decline until in later life the body often produces no progesterone.

Not only do oestrogen and progesterone work together to help reduce side effects, one of the most significant benefits of progesterone replacement is the prevention and treatment of osteoporosis, the bone thinning disease that mainly affects older women.

Progesterone can be applied to the face, neck, upper chest, breasts, behind knees, inner arms or inner thighs; the site should be rotated daily. Doses are usually around 12 mg to 24 mg. Pre-menopausal women would start the treatment on day 12 of the menopause and stop on day 25, repeating the cycle the following month.

10.6 GONADOTROPIN-RELEASING HORMONE (GNRH)

It is decapeptide hormone secreted by hypothalamus and controls FSH and LH release from adenohypophysis. GnRH is released in pulses, its pulse frequency and amplitude determines FSH : LH ratio. Pharmaceutical GnRH is synthetic and has analogues *Leuprolide, Naferelin, Goserelin,* and *Histrelin.* Analogues are more potent because of their longer half-life (reduced degradation) and higher affinity for receptor.

Antagonist of GnRH is available in the form of *Ganirelix* that has high affinity for GnRH receptor.

- *Leuprolide* endometriosis, fibroids, prostate, central precocious puberty (CPP)

- *Nafarelin* endometriosis, CPP

- *Goserelin* endometriosis, breast cancer, dysfunctional uterine bleeding, prostate cancer

10.6.1 Toxicology of GnRH

- Pituitary apoplexy and blindness if pituitary tumours exist.

- In women, it cause hot flushes, depression, decreased libido, generalized pain, vaginal dryness, breast atrophy, and osteoporosis (prolonged use).

- In men, it causes transient elevation of testosterone leading to pain for men with bone metastasis, hot flushes, edema, gynecomastia, and decreased libido.

10.7 HORMONAL REGULATION OF GENES TO TREAT DISEASE

Hollis-Eden is developing a series of potent hormones and hormone analogs that it believes are key components of the body's natural regulatory system. These Immune Regulating Hormones (IRHs) appear to be critically involved in controlling the immune system and metabolic functions (figure 10.1). Unfortunately, levels of these hormones are depleted as we age, and this process is accelerated as a result of chronic infectious diseases, stress or trauma. Preclinical and early clinical studies with these compounds indicate that they have the ability to significantly reduce a number of well-known inflammatory mediators, while also increasing innate and adaptive immunity and reversing bone marrow suppression. In addition, these compounds have a very attractive safety profile to date, are cost-effective to manufacture, and are unlikely to produce resistance.

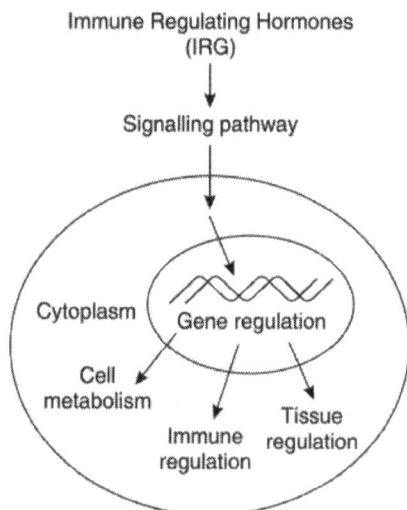

Figure 10.1   Mechanism of action of immune regulating hormones.

IRHs are based on naturally occurring human steroid hormones derived from the precursor class of hormones known as androstenes, which in turn are derived from the precursor pregnenalone. Numerous highly successful drugs have been produced from pregnenalone-derived hormones. Among these are cortisol and other corticosteroids that come from the glucocorticoid class of hormones.

Corticosteroids are used to treat a variety of ailments, including asthma and other inflammatory conditions. In many cases, however, use of corticosteroids is limited by their toxicity and immune suppressive effects. In contrast, IRHs appear from preclinical studies to share the anti-inflammatory properties of glucocorticoids, potentially without accompanying toxicity. As a result, IRHs may represent a class of "smart steroids" for use in autoimmune, inflammatory and other immune dysregulated conditions.

The androstene class of hormone precursors from which Hollis-Eden's immune regulating hormones are derived also has yielded significant and widely used pharmaceuticals. Among these are drugs derived from the hormone estrogen and prescribed for birth control and the treatment of osteoporosis and menopause. Drugs derived from the hormone testosterone are used in treating sexual dysfunction. Unlike the hormones estrogen and testosterone, however, the specific class of androstene-derived steroid hormones being developed by Hollis-Eden as IRHs are designed to minimize interactions with sex steroid receptors.

IRHs have demonstrated in either preclinical or clinical studies three important properties that distinguish them as an important and unique class of molecules—they appear to regulate innate and adaptive immunity, reduce inflammation and stimulate cell proliferation in the bone marrow. Each of these properties has significant beneficial utility in a broad array of diseases and disorders.

10.8 IMMUNOREGULATORY PROPERTIES OF CORTICOSTEROIDS

Glucocorticoid hormones (GCH) induce apoptosis in PHA (Phytohaemagglutinine)-primed peripheral blood T lymphocytes (PBL) and down-regulate membrane-bound proteins involved in the immune response. Researchers have analysed whether GCH are able to affect the expression of the TCR-associated molecules CD3, CD4, and CD8 on PBL-PHA, and whether the modulation of those receptors is related to the GCH-driven apoptosis of the PBL-PHA. Lymphocytes were cultured with PHA or with PHA plus prednisone (PDN) 10^{-3}, 10^{-6}, and 10^{-9} M. Then expression of CD2, CD3, CD4, CD8, and CD56 antigens was studied by cytofluorimetric assay using propidium iodide (PI) staining and annexin procedure, and by gel electrophoresis of low molecular weight DNA. PDN, at a pharmacological concentration (10^{-6} M), was able to inhibit the CD3 expression on T cells. The kinetics of CD3 decrement and of apoptosis show that the down-regulation of CD3 molecules precedes DNA fragmentation and that the cells lacking CD3 are those prone to PDN-induced apoptosis. The inhibition of CD3 is not related to a transcriptional or post-transcriptional phenomenon, because both PBL-PHA and PBL-PHA-PDN expressed the same amount of intracytoplasmic CD3 molecule. PDN also induced a down-regulation of the CD4 and CD8 molecules that resulted sooner in more intense CD8. *In vitro* PDN is able to induce apoptosis in PBL-PHA through a down-regulation of CD3 molecules.

10.9 GROWTH HORMONE (GH) AND BODY FAT

Somatostatin or growth hormone is secreted by anterior pituitary. It stimulates the growth and development of tissues, bones and muscle of body (see details in chapter 2). It is important to recognize how our muscle-fat ratio tends to change with age. Some 80% of a young adult's weight is lean body, mass-muscles, organs and bones. Only 20% is fat (adipose tissue). After the age of thirty, muscles atrophy partly due to genetic make-up and partly due to lifestyle, such as poor diet and lack of exercise. Every decade thereafter an average of 5% of lean body mass is replaced by fatty tissue, so that by the time most of us reach 70 yrs, we have gone from an 80–20 lean-fat ratio to a ratio that is closer to 50–50. Growth Hormone decreases 14% every decade as we age so that a 70-year old secretes only 20% of the HGH (human growth hormone) secreted by a 20-year old.

Dr Danial Rudman published a landmark study in the New England Journal of Medicine in 1990 when he showed that aging adults can reverse the physical signs of aging by as much as 20 years from injections of growth hormones. The most significant results were the drastic change in the ratio of muscle to fat. A study at the same institution after Dr. Rudman's death was published in 1996 and revealed the following, when 202 patients were given a low dose of growth hormone injection for several months:

- 88% reported improvement in Muscle Strength
- 81% reported improvement in Muscle Size
- 71% reported improvement in Body Fat
- 81% reported improvement in Exercise Tolerance
- 83% reported improvement in Exercise Endurance
- 71% reported improvement in Skin Texture
- 68% reported improvement in Skin Thickness
- 71% reported improvement in Skin Elasticity
- 51% reported improvement in Wrinkle Disappearance
- 38% reported improvement in New Hair Growth
- 55% reported improvement in Healing of Old Injuries

- 53% reported improvement in Back Flexibility

- 73% reported improvement to Common Illness

- 62% reported improvement in Duration of Penile Erection

- 57% reported improvement in Frequency of Nighttime Urination

- 57% reported improvement in Hot Flashes

- 84% reported improvement in Energy Level

- 62% reported improvement in Memory

Growth Hormone is safe but expensive. The combination of certain amino acids like (lysine, arginine, glutamine and ornithine) will spur the pituitary gland to release growth hormone especially when combined with the balanced diet and a good exercise program.

10.10 GROWTH HORMONE (GH)

In children, hyposecretion of growth hormone produces mental dwarfism while in adult low secretion of GH leads to Simmond's disease (dry and wrinkled skin with early old age signs and degenerated sex organs). If the lack of GH is diagnosed early enough, such children can be given extra GH to maintain their growth. By cloning the human gene for GH into a bacterium, the hormone can be produced from transgenic bacterial fermentation. A single 500-litre vessel filled with genetically engineered bacteria can provide GH equivalent to no less than 35000 human pituitary glands. In 1981, in the UK, 800 children could have benefited from treatment. In the US, about 2600 children are receiving GH. One child in 5000 suffers from hypopituitary dwarfism and the easy availability of this biopharmaceutical will be of immense benefit to these child sufferers. It is also evident that the GH hormone can increase muscle formation in normal individuals and is now being exploited by some athletes.

Eli Lilly and Company Ltd., one of the UK's top pharmaceutical companies, manufactured the human growth hormone, *Humatrope,* by recombinant DNA technology. The biological effects of Humatrope are similar to the human growth hormone made in the pituitary gland.

10.11 BIOTECHNOLOGY PHARMACEUTICALS

In US, biotechnology pharmaceuticals like Amgen, Bayer, Biogen, Elan, Enzon Roche/Genentech, Eli Lilly, Merck, MedImmune, Merch and Company, Johnson & Johnson, Novartis International, Schering-Plough, Glaxo Wellcome, SmithKline Beecham, Bristol-Myers Squibb, Pfizer, and Novo Nordisk are successfully involved in manufacturing hormones as pharmaceuticals and in India several multinational pharmaceuticals are not lagging behind in this field. With the advent of biotechnology, it is easy to produce pharmaceutical hormones like insulin, growth hormone and others on commercial basis from genetically modified organisms (GMOs). It is quite relevant to explain here the biotechnological approach to manufacture the human insulin from GMOs.

10.11.1 Insulin

It is a hypoglycemic hormone secreted by beta cells in islets of Langerhans of the pancreas (see details in unit 5.2). The deficiency of insulin causes a disorder known as diabetic mellitus. Until the advent of genetic engineering all the insulin given to diabetics was extracted from the pancreas of cattle or pig. The insulin molecule is made up 51 amino acids; the insulin has a different amino acid at one point as compared to human version, where as cattle insulin differs at three points. These small variation means that some patients eventually suffer allergic reactions to the human insulin. To avoid this contractions and considering the ever increasing number of diabetic patients, biotechnology made it possible to produce safe, efficient, low cost human insulin (Humulin) on a large scale from genetically engineered bacteria. Recombinant human insulin appears not to have any side effects and is increasingly having the largest market share of sale. Production is unlimited and free from market shortage of animals and all the problems associated with previous production methods.

10.12 GENETIC ENGINEERING FOR COMMERCIAL PRODUCTION OF INSULIN

As stated earlier genes form the fundamental basis of all life, determine the properties of all forms of life and are defined segments of DNA. Recombinant DNA technique, popularly known as gene cloning or genetic engineering, offers potentially unlimited opportunities for creating new combinations of genes that do not exist under natural conditions. Genetic engineering involves the formation of new combinations of heritable material by the insertion of nucleic acid molecule, produced by whatever means outside the cell, into any virus, bacterial plasmid or other vector system so as to allow incorporation into a host organism (it may be a bacterium, yeast, plant or animal cell) in which they do not naturally occur but are capable of continued multiplication.

Genetic engineering permits the selective alteration of the DNA in bacterial cells so that new genes are added to the DNA of the bacteria, with the result that the altered cells can synthesize nonbacterial protein like insulin, interferon, vaccines, growth hormone and antibiotics which are of immense biomedical or economic importance. It is possible to obtain an unlimited and chemically pure supply of these proteins with the aid of genetically engineered microbes. These techniques are useful not only in the study of basic processes in developmental biology (differentiation) but also in the actual alleviation of disease by gene transfer. In addition to this, rDNA technology is useful in the generation of new strains of crop plants (transgenic plants) and animals (transgenic animals) having altered characteristic features directed towards human welfare.

Recombinant technology has widespread applications, which not only involve the insertion and reproduction of eukaryotic genes in prokaryotic bacteria, but also the insertion, reproduction and expression of recombinant genes in eukaryotic cells and whole organisms. The various steps involved in genetic engineering are explained in brief with reference to manufacture of human insulin as follows.

10.12.1 Identification and Isolation of Insulin Gene

In the laboratory, the first step towards creating a genetically altered microbe, which will manufacture a human protein, is to identify and isolate the gene from a human cell that codes for insulin. With a few exceptions all cells of an organism contain the same genetic information, but they all do not use that information in the same way. The cells of different organs have different functions to perform, and this specialisation among cells is because of gene expression. This difference in expression between genes in specific cell types is used by genetic engineers as the most elegant method of isolating the required gene. This technique focuses on the cells' mRNA molecule, rather than searching through huge complex mass of genes encoded in DNA. Thus, if a gene to make insulin is required, human pancreas cells are examined to find mRNA molecule, which have been transcribed from that gene.

First the mRNA molecule must be extracted from the pancreatic cells. When these cells are broken open, they release materials including not only mRNAs but also enzymes, other proteins, DNA and pieces of cell membrane. All these materials are separated from each other by treating the mixture

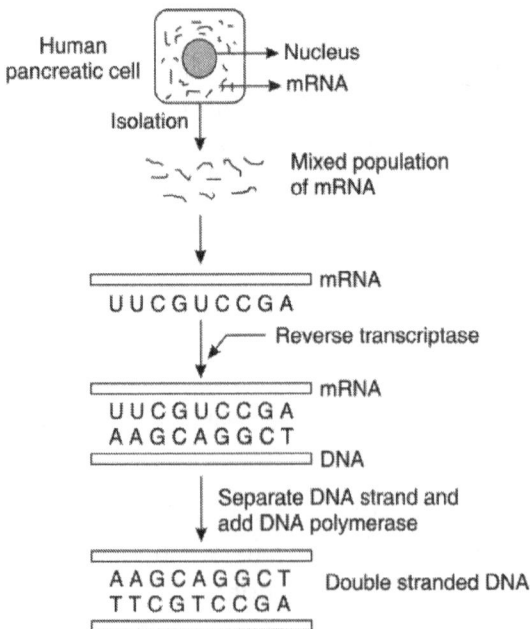

Figure 10.2   Process of formation of cDNA from mRNA by the enzyme reverse transcriptase.

with chemicals and subjecting it to a variety of physical forces. By centrifugation at various speeds and timing, different kinds of molecules are separated according to their weights and shapes.

The isolated population of mRNA molecules (mixed/ heterogeneous) contains the genetic information needed to manufacture all the proteins in the form of codons. This information is converted back into the form of the equivalent DNA sequences. With the help of a special enzyme, reverse transcriptase (RNA dependent DNA polymerase), it is possible to synthesise new single-stranded DNA molecule. This enzyme assembles the nucleotides in a linear sequence according to the order of bases on the mRNA, thus it uses mRNA as a 'template'. After a single strand DNA molecule has been copied from isolated mRNA, a complementary strand of DNA is produced using the standard DNA copying enzyme, DNA polymerase. Thus a new population of double-stranded 'copy' DNA (cDNA) molecules is produced that codes for mRNA that were in differentiated tissue. In other words, cDNA contains the same genetic information as the mRNA from which it has been formed. This cDNA is a copy of human gene (figure 10.2). Since the process started with many different types of mRNA isolated from human pancreas cell, there are now cDNA copies of many different genes among which are copies of the insulin gene.

10.12.2 Integration of cDNA into the Plasmid (vector)

Plasmids are small, circular extra-chromosomal DNA present inside the bacteria. They carry genes which enable bacteria to resist antibiotics. Plasmids have an independent ability of replication and they often pass from one cell to other even though the cells are of different species. These properties are most important for genetic engineering. Plasmids are used as vectors and are taken from one set of bacteria. The human cDNA gene is stitched into a plasmid circle and this is allowed to enter bacteria and transfer the human gene into its new home. The plasmid is cut open with a restriction enzyme that has a unique target site located in the sequence where the cDNA is to be inserted. *Bam*HI is the restriction enzyme used to split open the plasmid ring, leaving a set of 4 unpaired bases at each end of the molecule as shown in figure 10.3. The same restriction enzyme is used to clear the cDNA but before the action of the restriction enzyme, at each end of the cDNA, six linker bases are added. The restriction enzyme clears the

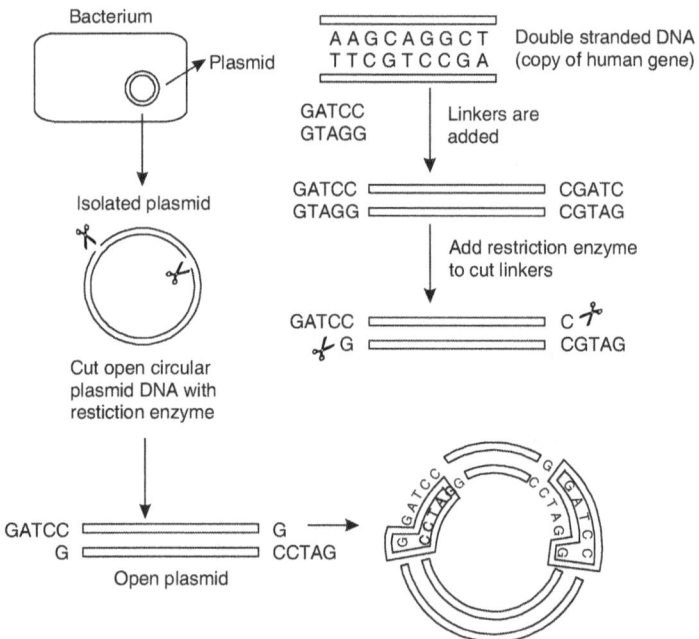

Figure 10.3 Method of preparation of recombinant DNA molecule.

cDNA through the linkers, producing unpaired bases that match those on the plasmid. The cDNA and the plasmid are then brought together to produce a single circular molecule. This is a recombinant DNA molecule or recombinant vector having integrated human gene cDNA. At this stage both the genes (cDNA and plasmid) are held together by 8 complementary weak bonds at their sticky ends. With the help of the enzyme DNA ligase, a permanent link is established between both the ends of genes providing stability to the recombinant plasmid.

10.12.3 Introduction of Recombinant Vector into a Suitable Host

Once *in vitro* construction of recombinant vector is completed, it is the bacteria that are to act as factories manufacturing the desired protein. The most popular choice of bacterial host is *Escherichia coli*. Plasmids have an inherent ability to enter the cells of *E. coli*, and adding a few simple chemicals to the mixture can facilitate this invasion. Once inside a bacterial cell, a single plasmid may multiply itself to produce a few dozen identical replicas. As the bacterium which harbours the recombinant plasmid is also growing and dividing as often as once every 20 minutes, each daughter cell takes with it a few of the plasmids, which again reproduce themselves (figure 10.4).

In a short span of time, a single bacterium will produce millions of descendants. A population of cells all derived from a single ancestor is known as a clone and all cells in a clone have the same genetic make-up. Thus a single bacterium carrying a recombinant vector will produce millions of identical cells, all of which contain the original human gene.

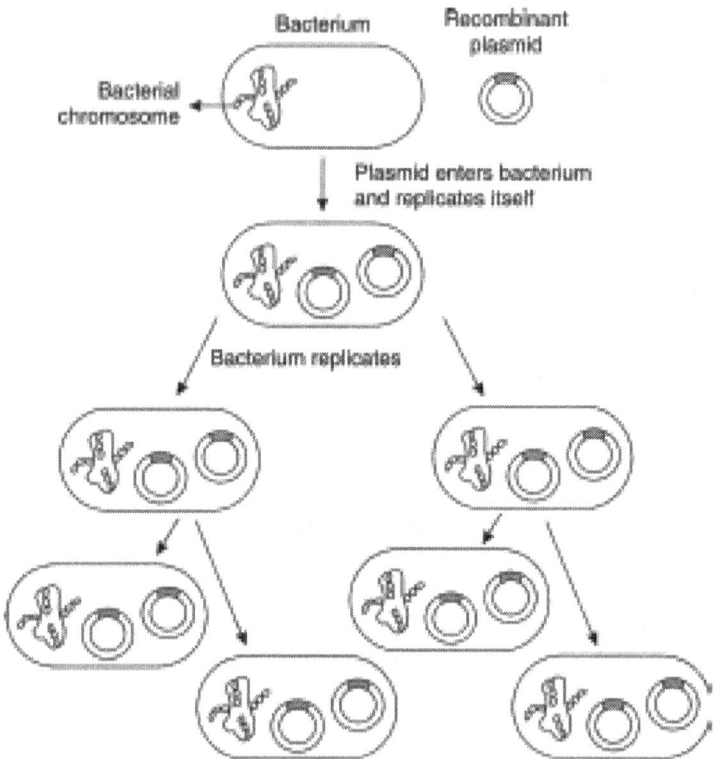

Figure 10.4   A cloning of recombinant plasmids containing human gene inside the bacterium.

10.12.4 Selection of Right Clone

The next task for genetic engineers is to make the right choice for the clones of the bacteria which contain a recombinant plasmid with its human gene. There can be 3 types of bacteria confronted by genetic engineers:

1. Bacteria infected by plasmids carrying a human gene (these are bacteria in which the scientists are interested).

2. Bacteria having normal plasmids (i.e. they do not have human gene).

3. Bacteria that have resisted the invasion by any type of plasmid (i.e. they are not transformed).

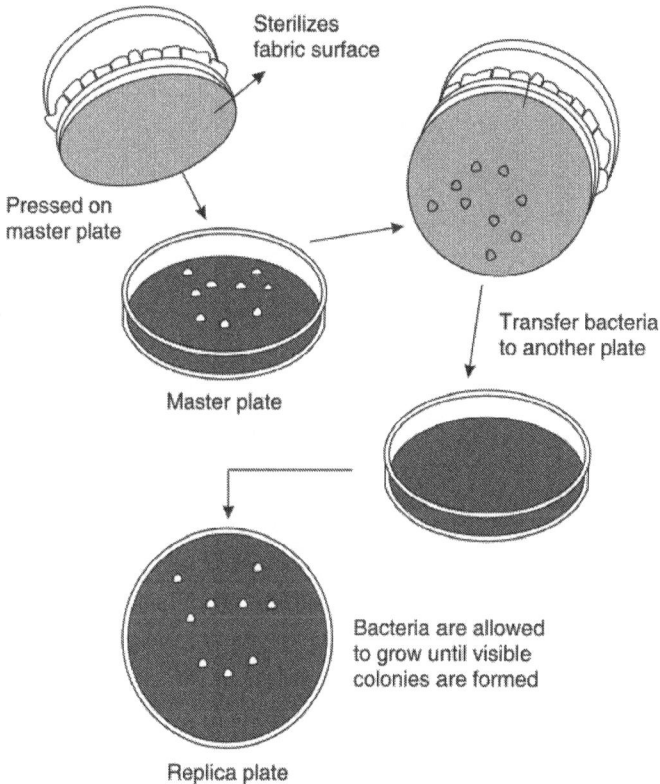

Figure 10.5 Preparation of replica plate with identical clones appear in exactly the same position.

The next step, therefore, is to identify the clone having recombinant vectors. Suitable selection strategies have been applied to achieve this objective, and this is the most important step in DNA cloning. A simple but invaluable technique that helps in the checking process is replica plating (figure 10.5). The mixed population of bacteria is smeared across a plate of nutrients to individual cells. Each cell is allowed to multiply until it has produced a visible clone of cells. Thus there will be

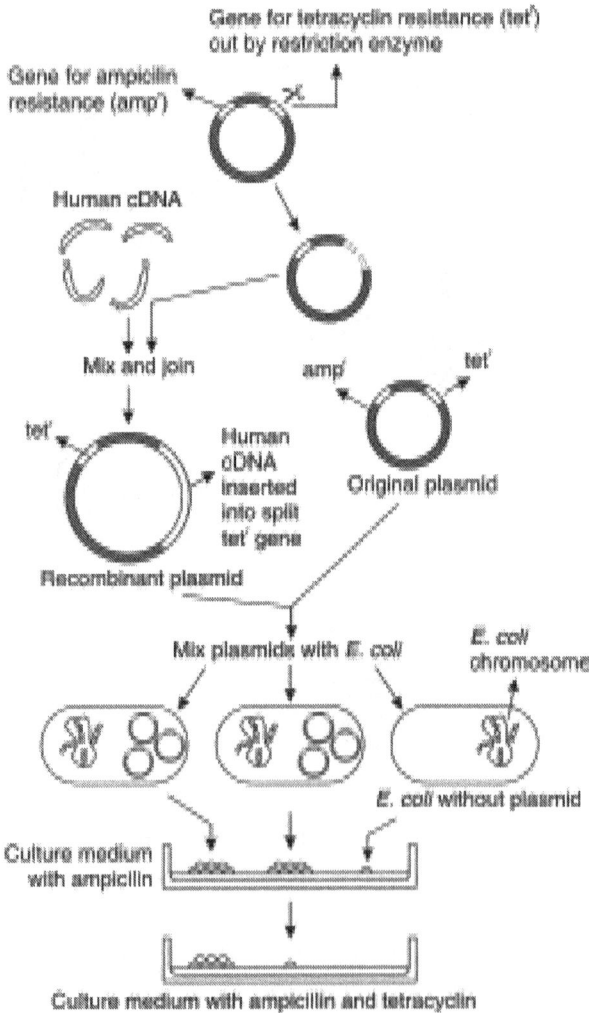

Figure 10.6   Identifying the bacteria which harbour recombinant plasmids containing human gene (cDNA molecule).

many distinct clones on the plate. Each clone is then separated into two portions, one is retained to grow in its ideal nutrients (master plate) while various experiments are carried out on the other portion (replica plate). Some of these experiments will kill some bacteria, but because their identical twins are kept alive and well in another place, every clone has copies of itself available for future investigations.

The bacteria which have no plasmids and those which have the normal plasmids need to be eliminated. Some types of plasmids carry genes that make bacteria resistant to certain antibiotics. One such plasmid is pBR322 which has the genes tet^r and amp^r for tetracycline and ampicillin resistance respectively. Any bacterium that contains the normal form of pBR322 will be unharmed by these antibiotics.

However, bacteria with recombinant plasmids will not be able to protect themselves against tetracycline since the introduction of a human gene in the middle of tetracycline gene has destroyed its protective action. This forms the basis of a test (figure 10.6) to distinguish between bacteria with no plasmids (these will be killed by either antibiotic present in culture medium), those that have the normal plasmid (these can survive in the culture medium containing both antibiotics) and bacteria that have recombinant plasmids (these resist ampicillin but succumb to tetracycline).

10.12.5 Colony Screening with Antibody for Pinpointing the Right Clone

After identifying the bacteria which harbour a recombinant plasmid/human gene DNA molecule, the next step is to search for the clones that manufacture the protein insulin. Since the whole procedure started with a mixed or heterogeneous collection of mRNAs (that code for many different proteins), only some clones will contain the gene responsible for production of insulin, while others contain different human genes. Radioactive antibody testing is one of the efficient and rapid methods for colony screening. The success of this method is due to two main facts—that antibodies can recognise and

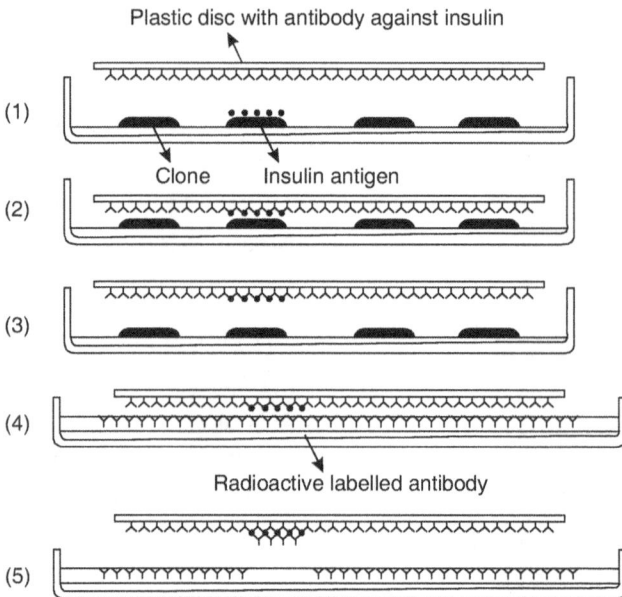

Figure 10.7   Pinpointing the right bacterial clone containing insulin DNA. A plastic disk coated with antibody to insulin is first exposed to the contents of cells from each clone.

   (1)   Any insulin present in the cells is bound to the antibody.

   (2)   And thereby fixed to the plastic disk

   (3)   Radioactive labelled antibody to insulin is then applied to the disk in order to detect the presence of the protein (4, 5).

attach onto specific types of proteins (antigens), and that tiny amounts of radiation can be detected.

Genetic engineers rely on the antibodies to search for bacterial clones that are producing insulin; by recognising insulin molecules, the antibodies stick to them like labels. These labels, like most molecules are too small to be seen directly. However, 'tagging' these labels with radioactive atoms helps them to become visible. These radioactively labelled antibodies emit very small amounts of radiation which produce a dark patch on photographic film. Figure 10.7 shows the combination of antibodies and radioactive labels that pinpoint to colonies of bacteria that manufacture insulin. Once insulin-producing bacterial clones are identified, they can be subcultured into an indefinite number of clones, producing a potentially unlimited supply of chemically pure, safe and effective insulin.

Biotechnologiests are trying to manufacture equally vital hormones including calcitonin, cholecystokinin, vasopressin, parathyroid hormone, nerve growth factor, adrenocorticotropic hormone, and erythropoietin from genetically engineered microbes.

SYNOPSIS

❑ In addition to implantation of *intrauterine devices* (IUDs), use of contraceptive pills made up of synthetic steroid hormones is the commonest and most effective method in human females to prevent pregnancy.

❑ *Contraceptive pills* are made up of an oral active oestrogen—*ethinyl estradiol* combined with a synthetic progestin—*norethindrone.*

- ❑ *Contraceptive injection* is also available in Indian market with trade name Depo-provera containing sterile *medroxy-progesterone acetate.*

- ❑ The progestin such as *levonorgestrel* is also used in the form of implants under skin to prevent conception.

- ❑ Prolonged use of synthetic oestrogen can lead to carcinoma in uterus and breast.

- ❑ The synthetic hormone *diethylstilbestrol* (DES) was prescribed to women to prevent miscarriage, to treat post-menopausal complications, to cure headaches, dizziness, nervousness, depression, frigidity, insomnia, muscle and joint pains, vaginitis, and infertility; and it was taken to prevent conception.

- ❑ *Hormone replacement therapy* (HRT) has the long-standing benefit of preventing bone loss in post-menopausal patients.

- ❑ There is a strong interaction between sex hormones and the immune system. The role of *oestrogen* in prevention of bone loss is mediated by mechanisms involving the inhibition of pro-inflammatory cytokines by bone marrow cells.

- ❑ Experimental evidences show a correlation between low DHEA (*Dehydroepiandrosterone*) levels and a declining immune system, and DHEA is being used in the fight against HIV, cancer and senile dementia.

- ❑ *Progesterone replacement* helps in the prevention and treatment of osteoporosis, the bone thinning disease that mainly affects older women.

- ❑ The synthetic GnRH has analogues *Leuprolide, Naferelin, Goserelin,* and *Histrelin,* which are more potent because of their longer half-life and are used to treat endometriosis, breast cancer, dysfunctional uterine bleeding, prostate cancer and central precocious puberty.

- ❑ *Immune Regulating Hormones* (IRHs) developed by Hollis-Eden help in controlling the immune system and metabolic functions.

❑ IRHs are based on naturally occurring human steroid hormones derived from the precursor class of hormones known as *androstenes*, which in turn are derived from the precursor *pregnenalone*.

❑ *Glucocorticoid hormones* (GCH) induce apoptosis in PHA-primed peripheral blood T lymphocytes (PBL) and down-regulate membrane-bound proteins (antigens) involved in the immune response.

❑ It has been proved experimentally that aging adults can reverse the physical signs of aging by as much as 20 years from injections of *growth hormones*.

❑ Most of the U.S. and U.K. based pharmaceutical companies flourish in their business by manufacturing the human growth hormone *Humatrope* and human insulin *Humulin* by rDNA technology from genetically modified organisms.

REVIEW QUESTIONS

1. Explain the term contraception. What are the different methods for preventing pregnancy? Describe various types of contraceptives.

2. What are the different side effects of prolonged use of synthetic steroid hormones?

3. Describe the term immune regulatory hormones. How are they involved in treatment of diseases?

4. Explain the pharmaceutical roles of corticosteroids with their immunoregulatory properties.

5. What is rDNA technology? How does it help in manufacture of synthetic hormones?

6. Write short notes on:

   i.   Intrauterine devices

   ii.   Synthetic GnRH

   iii.   Diethylstilbestrol (DES)

   iv.   Hormone replacement therapy

   v.   Humatrope

   vi.   Replica plating

   vii. Genetically modified organisms

GLOSSARY

Acidosis   A metabolic condition in which the capacity of the body to buffer H^+ is decreased and is usually associated with low blood pH.

Acromegaly   The disorder that arises in adults due to excessive secretion of growth hormone.

Adenylate cyclase   The enzyme that catalyses the formation of cyclic AMP from ATP.

Addison's disease   The clinical disorder produced due to primary adrenocortical insufficiency that usually occurs from an autoimmune process or from tuberculosis and other granulomatous diseases.

Aldosterone   One of the mineralocorticoids secreted by zona glomerulosa of adrenal cortex.

Amenorrhea   The condition in female where there are no menstrual periods.

Apoptosis   A programmed cell death in which the cell responds to certain signals by initiating a normal response that leads to the death of the cell.

Atenolol   One of the potent beta-blockers, commonly prescribed by physicians to block the β-effect of adrenaline on receptors on heart muscle cells, that helps to minimize cardiovascular disorders such as angina pectoris, hypertension, cardiac failure and myocardial infarction.

Autocrine secretion   The hormones that are self-stimulatory in action, that is, they act on the cells secreting them.

Barr body   Also called sex chromatin that is derived from X chromosome and present only in genetic female.

Beta-blockers   Drugs which selectively block β-adrenergic receptors, and can reduce the fight-or-flight responses to epinephrine. They are commonly prescribed by physicians to reduce stress.

Calcitonin   A peptide hormone secreted by parafollicular or C-cells in thyroid gland which acts to lower the concentration of calcium in blood.

Calcitriol (1, 25 dihydroxycholecalciferol)   The most active natural form of vitamin D.

Calmodulin   A calcium-binding protein, which activates certain enzymes like phosporylase kinase, adenylate cyclase, etc.

Catecholamines   The hormones like adrenaline and noradrenaline that are derived from amino acids such as tyrosine.

Cretinism   An infantile hypothyroidism in which the milestones of child development are delayed.

Cryptorchidism   A condition where either one or both testes remain undescended at any point along the reproductive tract.

Congenital nephrogenic diabetes insipidus (CNDI)   An inherited disease in which infants suffer from serious dehydration because of their inability to produce concentrated urine.

Conjugated protein   The class of protein that has additional prosthetic group made up of metal or other organic compounds.

Corticosteroids   The steroid hormones secreted by the adrenal cortex.

Cushing's syndrome   The disorder that results from excessive secretion of cortisol by the adrenal cortex or a cortical tumour.

Cyclic adenosine monophosphate (cyclic AMP, or cAMP)   A second messenger that is produced from ATP by an enzyme, adenylate cyclase, that is attached to the inside of the cell membrane.

Cyclic GMP   An analogue of cAMP, which can be formed by the action of guanylate cyclase enzyme from GTP (gaunosine triphosphate).

Depo-Provera   A trade name for the contraceptive injection containing sterile medroxy-progesterone acetate.

Diabetes insipidus   The physiological disorder arising due to lack of antidiuretic hormone. The condition is characterized by polyurea and polydipsia.

Diabetes mellitus   A relatively common disease caused by a deficiency in the secretion or action of insulin.

Down's syndrome (Mongolism or 21 Trisomy)   The disorder produced as a result of nondisjunction of autosome number 21. Persons with such abnormality are usually mentally retarded.

Dysmenorrhea   The painful menstruations characterized by severe menstrual cramps that are common in young women.

Dwarfism   The abnormal condition that occurs in children due to hyposecretion of growth hormone.

Ectopic pregnancy   An extrauterine pregnancy in which a fertilized ovum implants and begins to develop before it reaches its natural site in the uterus.

Effector A substance that brings about a cellular response to a signal.

Ethinyl estradiol An orally active synthetic oestrogen commercially available in the form of contraceptive pills.

Eunuchoidism The condition related to male hypogonadism dating from childhood, as a result of which the circulating level of androgens is depressed.

Exophthalmos The disorder produced as a result of hyperthyroidism that is clinically characterized by bulging of eyeballs.

Feedback loop A mechanism by which the output from a hormone's action feeds back into the system as input, switching off the hormone production once it has produced the desired effect.

Fertilin A protein present on the surface of the sperm head that helps in fusion of sperm to ovum membrane.

Gigantism A rare disorder that results from hypersecretion of GH in childhood and is characterized by overgrowth of all body tissues and organs particularly bones. The individual with this disorder may attain a height as much as 2.6 metres.

Glucose-tolerance test A sensitive diagnostic criterion used to define diabetes chemically.

Glycolysis The catabolic pathway in which a glucose molecule is broken down into two molecules of pyruvic acid.

Gluconeogenesis The biosynthesis of carbohydrates from noncarbohydrate sources like lipids and proteins.

G protein   A GTP-binding protein that acts as a switch to turn activities on or off.

Grave's disease or exophthalmic goitre   A disorder resulting from excessive secretion of thyroid hormones.

Gynecomastia   The condition in which breasts may be developed in male and may be found in patients with oestrogen-secreting tumour.

Half-life   A measure of the instability of a radioisotope, or equivalently, the time required for one-half of the radioactive substance to disintegrate.

Hashimoto's disease (lymphadenoid goitre)   A diffuse and massive lesion-causing goitre caused by hypothyroidism that is very common in middle-aged women.

Heteromeric G protein   A component of the transduction system known as G proteins because it binds guanine nucleotides (either GDP or GTP), and described as heteromeric because all of them have three different polypeptide subunits.

Hirsutism   The disorder related with excessive growth of body hair that results from increased secretion of testosterone or androstenedione.

Hormones   The special *chemical messengers* secreted by endocrine glands and that are carried to their respective target organ(s) through blood.

Hormone receptor   A protein molecule present on the surface of the target cells that binds to a specific hormone and triggers the cellular response.

Humatrope   A synthetic human growth hormone produced by recombinant DNA technology from genetically modified organisms.

Humulin   Human insulin that is commercially manufactured by several pharmaceutical companies by rDNA technology from genetically modified microbes.

Hydrocortisone   One of the glucocorticoids secreted by zona fasciculata of the adrenal cortex.

Hyaluronidase   The hydrolytic enzyme released by lysosome-like organelles in the acrosome of the sperm, which helps the sperm to enter the ovum during fertilization.

Hydrocele   The disorder in which a clear or straw-coloured fluid accumulates within the tunica vaginalis of the testes.

Hypercalcemia   It is a condition associated with elevated serum calcium due to primary hyperparathyroidism.

Hyperglycemia   Increased blood glucose level which is one of the symptoms of diabetes mellitus.

Idiopathic hypoparathyroidism   The rare acquired organ–specific autoimmune disorder that may be associated with primary adrenocortical atrophy, pernicious anaemia, chronic thyroiditis or thyrotoxicosis.

Intrauterine devices (IUDs)   Devices implanted into the uterus for prevention of pregnancy.

IGF-I   A growth factor structurally related to insulin that involves in responses of cells to growth hormone originally called somatomedin C.

Ketone bodies   Chemical compounds like acetoacetate, D-β-hydroxybutyrate, and acetone formed during partial oxidation of fatty acid.

Ketosis   A condition in which the ketone body concentration of the blood, tissue and urine is abnormally high.

Klinefelter's syndrome   A genetic abnormality where an individual has the genotype 44A + XXY and is characterized by mental retardation, underdevelopment of genitalia, and presence of feminine physical characteristics.

Ligand   Any molecule that can bind to a receptor due to its complementary surface.

Leuprolide   One of the potent synthetic GnRH used to treat endometriosis, breast cancer, and prostate cancer.

Male pseudohermaphroditism   The phenomenon that occurs in the genetic male whose testes are defective and who has developed external genitalia of female.

Melanocyte stimulating hormones (MSH)   The hormone secreted by pars intermedia of pituitary gland and that affects pigment dispersion in the melanophores of lower vertebrates, thus responsible for producing characteristic colour patterns.

Melatonin   A hormone secreted from serotonin by pineal gland or epiphysis and which regulates normal sleep patterns and shows antigonadotropic effect.

Motilin   The hormone secreted by stomach, small intestine and colon and acts on G protein-coupled receptors and causes contraction of smooth muscles in the stomach and intestine.

Myxoedema   A disorder produced due to primary hypothyroidism.

Negative feedback   A mechanism used extensively to regulate secretion of hormones in the hypothalamic–pituitary axis where an increase in the output of the system leads to a decrease in the input.

Nephrocalcinosis   The condition in which precipitation of calcium leads to formation of kidney stones.

Natriuresis   A condition where the excretory function of kidneys in relation to salt and water balance is altered due to the blocking of aldosterone (which is responsible for mineral metabolism) by progesterone.

Neurohormones   The hormones secreted by nerve tissue.

NO (nitric oxide)   The compound that acts as an intracellular messenger stimulating the cell's phagocytic activity. It helps in triggering the response leading to muscle relaxation and dilation of blood vessels with the help of cGMP.

Nondisjunction   A phenomenon in which a pair of chromosomes fails to separate, as a result of which both go to one of the daughter cells during gametogenesis leading to serious consequences.

Nontoxic goitre (simple goitre)   The non-inflammatory condition of the thyroid gland that results in enlargement of the thyroid gland without hyperthyroidism.

Osteomalacia   A disorder of bones that arises due to vitamin D deficiency in adults.

Oestrous cycle   The reproductive cycle exhibited by the females of all mammals except primates and includes the period of heat during which the female shows a desire for mating.

Paracrine secretion   Hormones which diffuse only short distances and act only on local cells.

Peptide hormones   Hormones that are chemically made of proteins or large polypeptides.

Phaeochromocytoma The diseased condition arising due to hypersecretion of catecholamines with a functioning tumour of the adrenal medulla.

Phospolipase C An enzyme that catalyses a reaction and splits PIP_2 into two molecules—1, 4, 5-triphosphate (IP_3) and diacylglycerol (DAG)—both of which play significant roles as second messengers in cell signaling.

Protein tyrosine kinase The enzyme that phosphorylates specific tyrosine residues of other proteins.

Receptor Any substance that can bind to a specific molecule (ligand), often leading to uptake or signal transduction.

Renal rickets A bone disease in children caused due to hyperparathyroidism.

Renin-angiotensin system A system that is present in juxta-glomerular cells of kidney helping in regulation of blood pressure along with aldosterone.

Restriction enzyme Enzyme that cleaves DNA duplex at highly specific sites.

Saheli A trade name for contraceptive pills that are developed by Central Drug Research Institute (CDRI) Lucknow.

Second messenger A substance that is released into the interior of the cell as the result of the binding of a first messenger, a hormone or a ligand, to a receptor at the outer surface of the cell.

Signal transduction The ability of the cell to convert an external stimulus to an appropriate cellular response.

Tetany   The disorder arising due to hypoparathyroidism in which the tone of skeletal muscles is increased with the spasm of the hands and feet.

Thyroid function test   The clinical evaluation of a patient for determination of hyper- and hypothyroidism.

Thyroglobulin   A colloidal material present in thyroid vesicles.

Thyrotoxicosis   The disorder resulting from excessive secretion of thyroid hormone.

Transcription factors   Auxiliary proteins that help in operation of eukaryotic polymerases.

True hermaphroditism   A condition in which a person has both ovaries and testes.

Turner's syndrome (gonadial dysgenesis)   A genetic condition where the genotype is 44A + XO and such individual is characterized by arrested genital development in juvenile state and no maturation at puberty.

Virilism   Appearance of secondary male features in female.

SUGGESTED READINGS

Anderson, J. R., (Editor), *Muir's Textbook of Pathology.* 11th Ed., 1982.

Cato, A. C. B., et al., "Regulation of gene expression by steroid hormones," *Prog. Nucleic Acid Res. Mol. Biol.,* 1992.

Cochran, B., "The Molecular Action of Platelet-Derived Growth Factor," In Adv. *Cancer Res.,* 1985.

Gerald Karp, *Cell and Molecular Biology: Concepts and Experiments,* John Wiley and Sons, Inc., 1996.

Goodman, H. M., *Handbook of Physiology,* Section 7: The Endocrine System, Oxford Univ. Press, 2000.

Guyton, A. C., *Physiology of the Human Body,* 5th Ed., Holt-Saunders International Editions, 1982.

John, B. H., *Clinical Diagnosis and Management by Laboratory Methods,* 17th Ed. W. B. Saunders Co., 1989.

Laurence, D. R. and Bennett, P. N., *Clinical Pharmacology,* Longman Group Publishers, 1987.

Marshall, C. J., et Al., "Reviews on signal transduction," *Cell,* 1995.

Martin, D. W., Mayer, P. A. and Rodwell, W. W., *Harper's Review of Biochemistry,* Lange Medical Publications, 1983.

Michael, S., *Hutchison's Clinical Methods,* 20th Ed. W. B. Saunders Co., 1995.

Neer, E. J., "Heteromeric G protein: Organizers of transmembrane signals," *Cell*, 1995.

Stryer, L., *Biochemistry*. Wiley International, 1992.

Watson, J. D. et al., *Molecular Biology of The Gene*, 4th ed., The Benjamin-Cummings Publishing Co. Inc., 1988.

INDEX

* 9 7 8 8 1 8 0 9 4 0 1 1 8 *